EPIPHANY Z

EPIPHANY Z

*8 Radical Visions for
Transforming Your Future*

THOMAS FREY

New York

EPIPHANY Z

8 Radical Visions for Transforming Your Future

Published in New York, New York, by Morgan James Publishing. Morgan James and The Entrepreneurial Publisher are trademarks of Morgan James, LLC.
www.MorganJamesPublishing.com

The Morgan James Speakers Group can bring authors to your live event. For more information or to book an event visit The Morgan James Speakers Group at www.TheMorganJamesSpeakersGroup.com.

Shelfie

A **free** eBook edition is available
with the purchase of this print book.

CLEARLY PRINT YOUR NAME ABOVE IN UPPER CASE

Instructions to claim your free eBook edition:
1. Download the Shelfie app for Android or iOS
2. Write your name in **UPPER CASE** above
3. Use the Shelfie app to submit a photo
4. Download your eBook to any device

ISBN 978-1-68350-017-9 paperback
ISBN 978-1-68350-018-6 eBook
ISBN 978-1-68350-019-3 hardcover
Library of Congress Control Number:
2016905180

Cover Design by:
Rachel Lopez
www.r2cdesign.com

Interior Design by:
Bonnie Bushman
The Whole Caboodle Graphic Design

In an effort to support local communities, raise awareness and funds, Morgan James Publishing donates a percentage of all book sales for the life of each book to

Habitat for Humanity Peninsula and Greater Williamsburg.

Get involved today! Visit
www.MorganJamesBuilds.com

DEDICATION

I'd like to dedicate this book to my beautiful wife Deb, who not only serves as President of DaVinci Institute, but also my right-hand person, traveling companion, sparring partner, sounding board for crazy ideas, all-around organizer of the unorganized, and manager of the grandkids and all social events.

vi | EPIPHANY Z

Deb and I first met when we were 13 years old, both freshmen at a tiny boarding school, Northwestern Lutheran Academy, in Mobridge, SD. After graduating together in 1972 we went our separate ways, only to find our paths inextricably reconnected in 2000, and forging an official you-and-me-against-the-world marriage pact in 2005.

Deb's role in DaVinci Institute should never be underestimated. Why my role has always been the big-picture visionary mapping out ideas in broad brush-strokes, Deb has been the implementer, the let's-make-this-happen person while carefully controlling the bottom line. None of this would have happened without her.

To our kids, Darby, Shandra, Kyler, Nicole, Jessica, and Bryan, who love listening to my half-baked theories, challenge my assumptions, and constantly keep me on task, I couldn't do this work without you. You and our 11 grandchildren serve as daily inspiration for everything that Deb and I do.

A special thank you to my niece, Hannah Frey, who created the graphics for this book. She is another family member who loves listening to my half-baked ideas.

CONTENTS

EPIPHANY Z

deas are the lifeblood of my existence. I often wake up in the middle of the night with a big idea, something I've dubbed the grand epiphany. But as it turns out, very few actually fit into the "grand" category.

Whenever they do, big ideas carry with them a heavy responsibility, the responsibility of either moving them forward or allowing them to die in the silent echo chambers of our own grey matter.

For this reason, I've often equated my eureka moments to that of being tortured by my own ideas. Yes, grand ideas are a wonderful playground where you can dream about starting a new company, solving some of the world's biggest problems, and constructing visions of wealth and influence, all in the time it takes most people to get ready for work.

However, what I've told you is far from a rare condition. Millions, perhaps even billions, are being equally tortured by their own epiphanies on a daily basis. In fact, every new product, book, movie, and mobile app has been born out of one of these lightning-strike moments.

Without epiphanies, life would be a monochrome experience. No swashes to add color to our dreams, no voices of urgency calling out in the middle of the night, and no moments of anticipation to cause our mind's fertile proving grounds to blossom. Instead, every silent box we open will be just that… silent.

I often weep for the great ones that have been lost. Every grand problem humanity faces today has been solved a million times over inside the minds of people unprepared to move them forward.

That's right, every major problem plaguing the world today, ranging from human trafficking to water shortages, major pollution issues, poverty, and even war has been solved again and again with personal epiphanies and no ability to advance them.

But that's about to change.

One of my recent grand epiphanies had to do with Moore's Law, the exponential doubling of capacity every 2 years.

With regular advancements in the physical world there is perhaps a 2X improvement every decade. But once an industry transitions into the digital world of the Moore's Law fast lane, it works like this:

- **Two Years—2X**
- **Four Years—4X**
- **Six Years—8X**
- **Eight Years—16X**
- **Ten Years—32X**

That's right, 32 times improvement in the digital world as opposed to maybe 2-4 times better in the physical world.

If we apply this to transportation, as the connected electric car goes digital, over the next decade they have the potential of improving by a factor of 32 while traditional mechanical cars only double.

As our physical houses enter the digital arena of smart homes, they too have the potential of improving by a factor of 32 while traditional homes only double in their improvements.

This also applies to our cities. A traditional city may double in improvements, while a smart city could potentially ratchet forward 32-fold over the same period.

Ironically, this exponential process also applies to epiphanies.

While most people end up being burdened with the responsibility of managing their ideas, I've figured out how to make them go digital.

For me, the process of dealing with every grand epiphany has been reduced to putting them into a presentation, or writing about them in a column.

My process has reduced the time from eureka moment to implementation to as little as 15 minutes. Ideas become the product, and rather than casting molds or bending metal, the way traditional products have been fashioned, a few images and descriptions form the basis on my digital products.

While this doesn't sound earthshattering, it truly is. This simplified process enables me to process perhaps as many as 32X the number of epiphanies in the time it takes the average person to deal with two.

This is a process I've termed—***Epiphany Z.***

Yes, if we had 32 letters in the English alphabet, I would have used the 32nd one. But Z will have to suffice.

If the past couple decades have taught us anything, it's been about our capacity for improving the world. My contention is that we now have far more capacity than what we are currently utilizing.

Over the coming chapters I will attempt to show how Epiphany Z thinking can be applied to virtually every aspect of our lives.

DESIGNING THE DASHBOARD OF YOUR DESTINY

I am a Futurist. Futurism isn't just part of my profession, it is a part of who I am. I spend my time studying, researching, thinking, writing, consulting and speaking about the future, a future that is coming into existence far faster than the second by second movement of the clock would indicate.

Anyone who has lived more than few years can see our present is quite different from even our recent past. If you've got more than a decade of years behind you, you've witnessed transformations—technological, economic, social, geopolitical—unimaginable to your parents when they held you as a newborn.

How do we best prepare ourselves—and our families, children, colleagues and co-workers and employees, our businesses and institutions, the world at large—for the dramatic transformations and disruptions that will accompany the enormous opportunities and possibilities the will present to us?

By learning how to think differently about the future.

Specifically, by developing the tools, skills and abilities—intellectual, intuitional, psychological, emotional—that enable us to *participate* in the development of the future coming our way, not simply to experience it, or, unfortunately for too many, let it wash over us or, sadly, let it wash us away.

Epiphany Z is my framework for developing those tools and skills and abilities, those *qualities* that will be necessary, even essential, if your dashboard is to *really* provide you with the information you need to control your destiny, or if it will only provide you with information about the forces controlling *you*.

And the first step toward taking control of your future is to re-think your relationship with the future.

We live in a very backward-looking society. We are backward-looking because we have all personally experienced the past. When we look around us, we see evidence of the past everywhere. All information is essentially history. The past is very knowable and comfortable.

Yet, we will be spending the rest of our lives in the future

From everything I've studied, past observers have focused on the concept of the future as being a consequence of time, rather than a stand-alone force of nature.

But—*is* the future merely the result of the movement of time, or *is the future itself a definable force?*

My raising of this question has led more than a few of my colleagues to think of me as that crazy guy who assigns human attributes to this thing we call the future. I don't think I'm crazy, but I will admit that on occasion you can hear me uttering phrases like, "I know it's going to be a great day because the future is clearly happy with me today." Or, "No, that's not a good idea because the future is probably going to push it off a cliff."

> "Is the future merely the result of the movement
> of time, or is the future itself a definable force?"

I don't really think the future is deliberately smiling upon me some days, frowning in my direction on others.

I *do* think the future is a definable force, and by studying the future we not only prepare ourselves to better handle the changes it brings our way, we also equip ourselves to, as the chapter title says, design the dashboard of our destiny.

Why is it so important to study the future? For starters, we all have a vested interest in it. We will all be living in the future.

12 Driving Forces Forging the World of Tomorrow

As a futurist, my job is to help people understand the driving forces forging the world of tomorrow. To make this more understandable, I've assembled a framework called the "12 Laws of the Future."

The best known example of laws pertaining to the future are the "Three Laws of the Future" written by Sir Arthur C. Clarke:

1. **When a distinguished but elderly scientist states that something is possible, he is almost certainly right. When he states that something is impossible, he is very probably wrong.**
2. **The only way of discovering the limits of the possible is to venture a little way past them into the impossible.**

3. **Any sufficiently advanced technology is indistinguishable from magic.**

Clarke often joked that Newton had three laws, and so three were also enough for him. Additionally, his friend and colleague Isaac Asimov, with whom he sometimes competed, had his Three Laws of Robotics.

However, in the 1999 edition of Profiles of the Future, Clarke added one additional law: "For every expert there is an equal and opposite expert."

Similar in some respects to what Clarke started, the laws I present here are framed around immutable truths that help us understand areas of science where little actual science currently exists.

12 Laws of the Future

1. **The future is one of nature's greatest forces.** The future is a force so massive the entire universe is being pulled forward in time simultaneously. We have no choice in this matter. *The future will happen whether or not we agree to participate.* There is no known way for us to speed it up, slow it down, or even try to stop it. The pace at which the future is unfolding is constant, and at the same time, relentless.

2. **The present is separated from the future by an invisible "field of knowability".** Everything in the past and present is knowable, but nothing about the future is completely knowable. We can personally witness, experience, and make sense of the present; but on the other side of this interface lies a veil of understanding we don't yet possess.

3. **Each of us experiences the unfolding of the future differently.** Every person is on his or her own personal journey. We each have our own ringside seat as we personally watch the field of

knowability reveal itself to us in a unique and different manner. We are the stars of our own hyper-individualized storyline.

4. **The future is non-existent until it passes the field of knowability, but we create our own approach vectors.** The energy that exists in the present creates an inertia that flows into the future. The inertia that is in place as we leave the present is still in place as we enter the future. If we witness the act of someone throwing a baseball, using a superfast strobe light, each billionth of a second of motion is tied directly to the next billionth of a second of the baseball's journey. Our inertias give motion to the present and direction to our future.

5. **The future is being formed against and within a backdrop of existing inertias.** On a personal level, we are all dealing with the inertia of our body and the inertia of our mind. Both are constantly in motion. At the same time, our personal inertias are taking place inside the context of every other person's inertia, as well as the inertia of every other thing around us. Nature has its own sets of inertias, with the forces of nature providing the inertia for every living and non-living molecule in the entire universe.

6. **Predictions are based on probabilities, and most of our future is being formed upon a foundation of stable slow-changing elements that can be predicted with a high degree of probability.** As humans, we tend to focus on the volatile, and ignore that which is stable. Buildings, trees, and mountains change very little from one day to the next. Only rarely do they undergo a radical transformation quickly. The earth's orbit around the sun, the speed of light, the changing of the seasons, the schedule of tides, the frequency of quartz crystals, and the laws of gravity are all predictable with a high degree of probability. Without having this assurance of

predictability, planning for the future, even near-term futures, would be very difficult.

7. **The future is not a human-centric force.** Without human influence, the future tends to be very cold and unforgiving. The future doesn't care whether you're happy or sad, employed or unemployed, married or single, personally content or emotionally adrift. The future is like a machine, impervious to our wishes, ambivalent to our goals. Only humans care about these things. At the same time, without humans, the future doesn't matter because there will be no one to care.

8. **Amidst this backdrop of existing inertias, the future is ours to create.** We do not have direct control over the future, but new inertias can be created, and existing ones can be influenced. The future is constantly being formed in the minds of people around us. Each person's understanding of what the future holds will influence the decisions they make today. As we alter someone's vision of the future, we also alter the way they make decisions in the present.

9. **Thinking about the future will cause it to change.** The very act of thinking about the future creates a new inertia, and this inertia changes our energy flows into the future. The "future part of the brain" is like a muscle that rarely gets exercised. But, like all muscles, the more we use it the better we get at leveraging the powers and energies of the future.

10. **The future is filled with power and energy.** The inertia of all matter in the universe is like a massive river of power and energy flowing from the present into the future. As humans, we only have the ability to affect a tiny microcosm of change. But our seemingly insignificant existence can have massive implications.

11. **Every avalanche begins with the movement of a single snowflake.** Our ability to tap into and leverage the power of

the future is directly tied to the number of times we think about it. The more we think about the future, the more we expand our understanding of it. And the more we understand the future, the easier it becomes for us to interact with it.

12. **The "unknowability" of the future is what gives us our drive and motivation.** The fact that the future is unknowable is a good thing. Our involvement in the game of life is based on our notion that we as individuals can make a difference. If we somehow remove the mystery of what results our actions will have, we also dismantle our individual drives and motivations for moving forward.

My goal in presenting these "laws of the future" is to prompt a conversation aimed at further refining our thinking.

One key to this refinement is a theory I call:

The "Field of Knowability"

I begin with the assumption that there is a small gap in time between the point when the future is formed, and the point when we know about it.

That point when we consciously become "aware of the present" is something I referred to briefly in the Laws, and which I'll now spend a little time explaining in more detail.

In my hypothesis, the "present" exists for a tiny period of time, a fraction of a second, before we ultimately experience it. Think of this as a staging area for what occurs next.

Hoping to understand more about the dividing line between the present and the future, I constantly ask:

- **When does the future end and the present begin?**
- **How does the future become 'now,' and where does it go from here?**

When talking of time and the future, it's all too easy to slip into thoughts about premonitions, ESP, and similar unexplained phenomena; such matters have no place here. I'm searching for a hard-science approach to the unveiling of the present.

Reality is constantly unfolding, but is it also possible there is a time-delay between the formation of the present and the moment when our human senses are able to comprehend it? Such a time-delay would help explain many of today's mysteries of science and physics..

As an example, the nature of gravity has remained one of mankind's greatest unsolved mysteries. Suppose the forces of gravity were exerted in this ultra-brief period of time, prior to knowability, with the only thing remaining once the present arrives, being the *effects* of gravity, not the measurable force itself.

How can we create an experiment to test this hypothesis? The first test needs to prove the existence of a "present" before we know about its existence, even if it's only a fraction of a second before human knowability emerges. The second test would be designed to pierce the field so we can "know before we know."

In short:

How do we time-shift the present into the pre-present?

Separating the Present from the Future

Einstein described it this way.

> *"The distinction between the past, present and future is only a stubbornly persistent illusion."*

The present is constantly forming around us: in many respects, we are swimming in the present. And like fish immersed in water, we have no way of gaining an outsider's perspective of something we are continually submerged in.

The movement or progression we experience from one moment of time to the next is driven by what I call inertia. The inertia in place as we leave the present is still in place as we enter the future.

Our inertias give motion to the present and direction to our future.

Each of us deals with the inertia of our body and the inertia of our mind. Both are constantly in motion.

At the same time, our personal inertias take place inside the context of every other person's inertia, as well as the inertia of every other thing around us.

Nature has its own sets of inertia, with the forces of nature providing the inertia for every living and every non-living molecule in the entire universe.

Unlike those non-living molecules, we possess the ability to generate ideas. And it's in those ideas that we begin to see the future take shape, even while we're still on this side of the Field of Knowability.

Which is why it's worthwhile to think about ideas and where they come from.

And even more worthwhile to think about where they go.

Where do Great Ideas Come From? Where do They Go?

Recently my wife Deb came up with a rather comical phrase to describe her occasional memory lapse, referring to her "photogeriactric memory."

I started thinking about where original ideas come from, and whether or not this was truly an original idea.

After a few Google searches, I found a total of 83 results for the term "photogeriactric" and an obscure 2007 reference to the phrase "photo-geriactric memory." So Deb's phrase wasn't totally original, but the question of originality continued to fascinate me.

Is the arrival of a creative idea a form of:

- **Manifest destiny (right place right time)**
- **Divine providence (this idea was destined for you no matter what)**
- **Sheer intuition (resulting from your connection with here and now)**
- **Something more like a ripple in the force (bubbling up from the idea cauldron of life)**
- **All of the above, depending upon the context**

Original ideas can be tremendously valuable, and if we know where they come from, we should be better equipped to create more of them.

Each of us is an idea-generating machine. We radiate ideas the way the sun radiates light.

It's true that, like Deb's phrase, very few are truly original ideas. But some are. We now have over 7 billion people radiating ideas, every second of every day, casting pieces of inspiration and brilliance in every possible direction. The question we should be focusing on is: where do they go?

Meet the Idea Expert

A couple of years ago I had a chance to meet Steven Johnson, author of the *Where Good Ideas Come From*. Johnson's book focuses on Eureka moments, the imagination sparks and the *epiphanies* where brilliant insights suddenly appear in our heads.

Steven concluded that people's so called "moment of inspiration" usually took place over an extended period of time, rarely happening all at once. As he describes it, most discoveries start with a "slow hunch" that builds through a series of micro-epiphanies until the entire breakthrough puzzle has all its pieces in place.

During our short discussion, Steven mentioned that idea people often create their most fertile environments working with what he called "liquid networks". With the right collections of minds focused on a topic, a well-executed discussion can lead to synergistic mind-bolts that can quickly be fashioned into completely functional ideas.

When I started ranking epiphanies on a comparative scale. I quickly found that it's hard to understand the significance of an idea when it first occurs. It's equally hard to uncover early what are later revealed to be the idea's fatal flaws.

Most of our ideas tend to be little ideas, micro-epiphanies. But I have also found that in every cluster of micro-epiphanies there is likely to be one with marquee-flashing-headline potential.

And when that headline idea occurs the question always becomes, "Now what?"

Creating a Fertile Growing Environment for Epiphanies

Whenever we form a great idea, we look for a place to put it. Ideas, much like parasites, need a host. If we don't manage to gaff them before we slip into our next stream of consciousness, they will be forever lost. Without a host, these squirming little idea-fish have a very limited shelf life.

If we manage to cluster enough of them together, they acquire a bit more staying power. But the cluster of ideas must somehow reach critical mass before they become noteworthy.

In the past we had very few options for pursuing that critical mass. We could jot ideas down in a notebook, mention them to friends, or make a few drawings or sketches. But even after such actions, most ideas died of isolation. We had very few "places" to put our flashes of brilliance.

Today our options have grown exponentially—good ideas can go from zero to Facebook entry in 0.9 seconds. They can be fashioned into tweets, infographics, photos, podcasts, PowerPoints, LinkedIn discussions, Quora forums, YouTube videos, submitted to blogs, or turned into interactive charticles.

"Ideas, much like parasites, need a host."

New platforms for sharing ideas arrive almost daily. Tomorrow those options may be near-infinite

We literally have thousands of placeholders for our momentary flashes of brilliance. Much like planting seeds in the freshness of damp soil, these memes have the organic potential to spring to life and ultimately be harvested as a colorful bouquet.

Creating a Picture of the Bigger Picture

Every social network, discussion forum, or live webcast has become a cosmic breeding ground for Steven Johnson's "liquid networks"—a place where ideas often have sex with other ideas.

Ideas have a way of creating structures in our minds, and these structures become self-assembling and self-constructing, filling in the gaps, adding multi-dimensions to our usual flat way of thinking.

Our future is being crafted by human genius in an organic sea where the best of the best find a way of rising to the top. We have seriously shortened the distance between problems and solutions, pain and comfort, and products and ideas.

The better we become at filtering the signal to noise ratio of human epiphanies, and leveraging these storehouses of ideas, the quicker we reach what's on the other side.

That's what EPIPHANY Z—the concept itself, and this book specifically—is all about. How to prepare ourselves—through learning and study, observation and reflection, conversation and debate—to thrive in the dramatically different future that exists on the other side of the "field of knowability," a concept that will continue to challenge you in the chapters ahead.

And not just to thrive—to take actual control of those forces of the future, the ones that touch us personally, and shape them to benefit ourselves as well as our fellow citizens of the world.

Each of us possesses the potential to do all that and more. Epiphany Z will show you how to transform that potential into the tools, skills, abilities, and above all the *vision* that lets you bridge the knowability gap and make the future—*your future, the future your dashboard will control*—a very real part of your living present.

Guided by Epiphany Z, you will see that the future is, very literally, an ecosystem composed of ideas. And in the chapters ahead I will show you how to make those ideas a potent ally in the years ahead.

EPIPHANY Z IS FAR LESS ABOUT WHERE IDEAS COME FROM AND FAR MORE ABOUT WHERE THEY GO AS THEY ENTER OUR EMERGING IDEA ECOSYSTEM!

EPIPHANY Z—
OPTIMIZING PURPOSE

H ow did we get here?

And how do we get from here to Epiphany Z?

I don't have all the answers—no one does,

But I do have some of them . . .

As a baby, life was all about eating, sleeping, and dry diapers. Even though you are learning new things quickly, not much else really mattered. By the time you enter grade school, you have learned to walk, talk, feed yourself, and have fun with your friends.

Entering high school you've grown much taller, in most cases, doubling your height from when you were a toddler. Your eyes and facial features have many similarities and look familiar, but you are now very different. You are fascinated by music, television, and any time you spot a passing smile by someone of the opposite sex, it becomes heart-stoppingly important. Relationships matter.

Every new day has you seeking a different set of experiences. You take pride in whatever you were good at, and become enamored with things you enjoy.

Every personal relationship brings with it a different set of involvements. Your first kiss sets the stage for your second, and your first intimate moments become cemented into the very fabric of your being.

As you enter your thirties and forties, your skillsets change dramatically. With age comes perspective, big problems become little ones, and over time, even the little ones fade away. In your sixties and seventies you begin to feel time is running out. It is in this progression we begin to realize that the future has changed us every step of the way.

Even though there are continuities to our personality and genetic structure, we are constantly changing. One cell gets replaced by another until we bear little resemblance to that person we were so many years ago.

And yes, you are now a different person than you were, even a few seconds ago.

Does that mean your past self is irrelevant—particularly in a future that's changing at the speed of light?

Hardly!

18 Reasons Why the Person You Were Still Matters

The former you has set the stage for the present you, and the person you are today will become critically important to the person you become in the future.

1. **Memories**—Every past memory helps crystalize who you are today.
2. **Shared Experiences**—Every long-term relationship is built around shared experiences, and these shared experiences provide the common ground foundation for future ones.

3. **Emotional Values**—Everything around you is constantly being emotionally rated on a subconscious level. That is why your car will generally hold more value than things like a skateboard or power drill.

4. **Skills**—Learning how to perform a task efficiently ties directly into a combination of short-term, long-term, and muscle memory. While some skills will fade over time, their influence will remain for years to come.

5. **Your Body**—Your present body came from your former body.

6. **Derivative Talents**—Every talent you have is a derivative of some other talent, interest, or tendency.

7. **Physical Improvements and Physical Impairments**—Every time you work out, it causes both short and long-term changes to your body and health. On the flip side, every time you hurt or injure yourself, it will also cause residual effects that linger over time.

8. **The Personality Equation**—Every individual is a combination of attributes, tendencies, desires, interests, and about twenty more ingredients we don't have names for yet. Some will change significantly over time, but others less so.

9. **Secrets**—Hidden deep beneath the sub-floor of human consciousness are our secrets that can come back to haunt us if we don't deal with them somewhere along the way.

10. **Struggles**—Our struggles are what make our accomplishments valuable.

11. **Obsession**—Determination becomes obsession and then it becomes all that matters. But from my vantage point, obsession is underrated.

12. **Possessions**—Yes, it is possible to simply walk away from all of our possessions, but few people do. Not only do we own our

possessions, they own us. And the things we own very often influences our future decisions.

13. **Connections & Networks**—We forge our weak and strong relationships through our connections. But today's social networks give us the tools to amplify those connections in a massively powerful way.

14. **Inner Voice**—Our most intimate of all intimate relationships takes place in the rarely audible space inside our head. We have a constant love-hate relationship with our inner voice, and even though we argue with ourselves, it will continue to influence who you are in the future. No it won't! Y.E.S., I.T. W.I.L.L.!

15. **Hopes & Desires**—Inside every great person is the hope and aspiration to become something better—more meaningful, more influential, more passionate.

16. **Reputation**—If we're doing things correctly our reputation will enter the room before we do. Our reputation involves a multitude of variables, and is one of the most influential aspects of who we are.

17. **Quirkiness**—Todays foibles can become tomorrow's most admired qualities if we know how the leverage them.

18. **Legacy**—For many of us, the disturbance we leave in the force field of life is the most significant accomplishment we can possibly make.

18 Reasons Why the Person You Were No Longer Matters

The voice of the fatalist inside often gives us little room for hope. If we believe change is not possible, then it certainly isn't.

1. **You look different.** You're nearly unrecognizable to those you hung out with twenty years ago.

2. **You've forgotten.** The vast majority of your life has disappeared into the ether, leaving little more than a faint residue of the imprint you made along the way.

3. **Your physical abilities have changed.**

4. **Your income is different.**

5. **Your friends are different.**

6. **Your clothes no longer fit, and if they still fit, they fit differently.**

7. **The things you valued most in the past, now holds little meaning.** (Note to self—Shag carpeting should have never been invented.)

8. **Your favorite sports team today has none of the same players you remember from ten years ago.**

9. **New friends may be more valuable than old friends.**

10. **Your ability to make brilliant decisions today is far greater than the person you left behind.**

11. **Past mistakes can only haunt you if you're still you.**

12. **Bad memories can be replaced by good ones, and old dreams can be replaced by more inspiring, more infectious, more exciting new dreams.**

13. **New skills will make you a different person.**

14. **Every significant shift in your life can be broken down into a series of baby steps that can be repeated, modified, redirected, or recalibrated.**

15. **You are only one relationship away from being the person you want to be.**

16. **There is always a path out of your current dilemma.**

17. **The only thing holding you back is you.**

18. **There is no limit to personal wisdom.**

Each morning, as I brush my teeth, I barely recognize the person in the mirror staring back at me. If I'm the same person I was twenty years ago, then why do I look so different, think so different, and why has my path of progress been so unpredictable?

Those are questions we all ask of ourselves when questioning or seeking to define our purpose.

But there are also bigger questions to be asked, ones that are beyond our normal everyday thinking.

10 Questions that Neither Science nor Religion Answer

As I was taking a tour of a dome-shaped house many years ago, the architect explained to me that domes are an optical illusion. Whenever someone enters a room, their eyes inadvertently glance up at the corners of the room to give them the contextual dimensions of the space they're in. He went on to explain that since domes have no corners, from the inside they appear larger than they really are, and from the outside they appear smaller than a more traditional house with a comparable footprint.

This notion of context has followed me throughout my life; into virtually every topic I've come to wrestle with. Once I find the "corners of the room" I can begin to make sense out of whatever subject I'm dealing with. However, when I dive into the "why" topics of how time and space began, and even the size of the universe, I find myself struggling to even formulate good questions. I can't find the "corners"—but I continue to look for them.

Perhaps this is nothing more than a form of therapy for me, but I'd like to take you along on an inner personal journey into how I think about the biggest of all big picture issues.

The journey starts with ten "simple" questions:

1.) *Why are there exceptions to every rule?*

Why is it all of our rules, theories, maxims, and models all have an exception? This is precisely the way the world works, except when it doesn't.

In a perfect world, we wouldn't have any exceptions—or would we? On the surface this seems like a rather trite question, to which most people would simply smile, shrug, and move on. But, in a world where scientists have spent countless billions to research and understand such topics as the relationship between matter, energy, particles, and waves, everything has to make sense, except it doesn't.

Why do exceptions matter?

Exceptions matter—nothing is 100% predictable. We can count on such things as buildings existing from one day to the next, the earth traveling around the sun in the same orbit, gravity holding us down, and the speed of light remaining reasonably constant. In fact, most of the world has been created around natural forces that can be predicted with high degrees of probability. But probability is *not* certainty. There is no such thing as absolute certainty, except our certainty that nothing is certain… maybe.

2.) *Why do logic and reason fail to explain that which is true?*

In many scientific circles, the only truths are those that can be explained with logic and reason. Religious people use a different metric, but they, too, have a way of calibrating their truths with logic and reason.

Why are logic and reason such miserable tools for explaining the world around us?

It's as if the world around us was perfect, and then someone divided by zero. Everything perfect has a touch of that one secret ingredient known as chaos.

Is order more perfect than chaos? Or is chaos just a higher form of order? How will we ever know if we can't explain it with logic and reason?

3.) Is the universe finite or infinite?

If we were able to travel to the outer edges of the universe, what would we find? Perhaps we would run smack dab into another universe, but how would we know?

Would the other universe somehow come in a different color, operate with a different set of rules, or smell slightly like almonds? How would we know?

I'm imagining a large sign that says, "You have reached the end of Universe A! Welcome to Universe B, where proximity is not an issue!" How much is infinity plus one?

4.) Why does anything exist?

Before there was something, there was nothing. And out of nothing, how did we get something? What existed before the big bang, before creation, and before God?

Yes, it becomes very confusing when we throw in theories about other dimensions and non-linear time, but all of these theories fail to answer this most fundamental of all questions, "Why does anything exist?" We know things exist, but why?

5.) Why does time exist?

Time is the sound of a metronome ticking in our heads, the beat of our heart, the blinking lids on our eyes, the mental waves in our brains, and all the circadian cycles that govern our lives. Much like fish that can't understand water because they're in it all the time, we have a very poor grasp of our most immersive of all substances—time.

Each of us thinks about time differently. To some it is a tool to be leveraged, to others a setting sun, a theory of physics, a philosophy to be debated, the hands of a clock, a lengthening of a shadow, or the grains of sand dropping in an hourglass.

And yet every truth we have about the existence of time comes with a counterbalancing exception to the rule.

I love Albert Einstein's comment that "the only reason for time is so that everything doesn't happen at once." What Einstein may have been alluding to is the existence of other dimensions outside of those governed by time.

But every time he made the comment, it always ended with a smile, the universal sign for "no further explanation will be forthcoming."

6.) Why do humans matter?

We are born as a baby, struggle our entire life with everything from finding food to eat, homes to live in, educating ourselves to gain more understanding, staying healthy, making friends and relationships, raising a family, earning a living, and then we die.

If we have more accomplishments in life, earn more money, have more friends, raise a bigger family, and somehow do everything better than anyone else, we will still eventually die. Right?

In a world teaming with 8.7 million different life forms, how do humans fit in?

Past civilizations, with their manmade structures, machines, systems, and cultures, has eventually succumbed to Mother Nature. Plants, animals, bacteria, and fungi methodically remove every trace of what we leave behind.

Why are humans important?

Does the fact that we can ask questions like these, ponder the un-ponderable, think the unthinkable, and accomplish things that no other species can accomplish, somehow give us a higher purpose? Are humans destined to become the guardians, caretakers, and eventually the masters of the universe?

If so, then we have to ask…

7.) *Why are humans so fallible?*

Humans are the bull in every china closet, the off-center bubble on every level, the mystery behind every hidden agenda, and the blunt instrument whenever a precision tool is called for.

We are both our greatest heroes and our most feared enemies. We are praised for our accomplishments and castigated for our failures.

Of all species on planet Earth, humans are the least predictable, most destructive, require the longest nurturing period, and consume the most food. At the same time, we are also the most curious, most aware, most innovative, and the most likely to waste countless hours playing video games.

Yes, we may have better developed brains than all the other animals, but that doesn't explain why we are so unbelievable fallible?

8.) *Do human accomplishments have long-term meaning?*

If you do a search of mankind's greatest accomplishments, you come up with lists that include the building of the great pyramids, landing on the moon, the invention of the telephone and light bulb, amazing artworks, and the composition of countless music scores. But are those things humans humans consider to be great accomplishments really significant in the bigger scheme of things?

Perhaps today's human accomplishments are a stepping-stone to what comes next?

We live in a world driven by prerequisites. A machinist needs to understand a single-point lathe operation before he or she can advance to multi-axial milling. Engineers need to understand the concepts of mechanical stress and strain before they start bending a cantilever beam. Metallurgists need to understand thermodynamics before they attempt phase transformations in solids. Physicists need to understand quantum mechanics before they can understand a standard model for particle physics. Mathematicians need to understand nonlinear differential equations before they can understand strange attractors.

Are all our accomplishments just stepping-stones to something else we don't know or understand yet? So what is it we don't currently know that will make tomorrow happen?

9.) *Why is the future unknowable?*

While I'm well aware of the notion that a "known future" will strip us of our drive and motivation, understanding the consequences still doesn't explain why the future isn't knowable. I like to think of the future as a force so massive the entire universe is being pulled forward in time simultaneously.

We have no choice in this matter. The future will happen whether or not we agree to participate. Currently there are no known techniques for us to speed it up, slow it down, or even try to stop it. The pace with which the future is unfolding is constant, and at the same time, relentless.

Will the future always remain unknowable?

10.) *What is the purpose of death?*

Shortly before his death, Steve Jobs said, "No one wants to die. Even people who want to go to heaven don't want to die to get there. And yet death is the destination we all share."

But why death?

Couldn't we just dissolve into a pile of ash, fly out of our skin, step into an invisible elevator preprogrammed to go to the highest of all floors, or just mentally fade to black. People fear death. We spend millions on vitamins, health food, fitness programs, and doctors all to avoid the unavoidable. Or is it unavoidable?

"The future will happen whether
or not we agree to participate."

The Feud Between Science and Religion

It's rather ironic that our first impulse is to use logic and reason to come up with answers, an approach that has historically only been able to answer questions about the tiniest of all fractions of the knowable universe.

Even before the time of Copernicus, scientists such as Philolaus and Aristarchus of Samos proposed something other than an Earth-centered universe.

While evidence of this line of thinking had been building for centuries, with Nicolaus Copernicus publishing his landmark book *On the Revolutions of the Celestial Spheres* in 1543, it wasn't until Galileo made his mark in 1615 that the rift between science and religion would reach death-sentencing proportions. However, there are some seemingly unanswerable questions that neither science nor religion can offer a reasonable answer.

People who surround us today are part of the present and will also be part of the future. For people who are intellectually enlightened and "tuned in," it's easy to discount those who have a different perspective. Yet the future is being created by all of us. If we believe we have a purpose, then so does every butterfly, pocket mouse, and beam of light.

Optimizing Evil

We have all experienced things the would consider extra-dimensional, such as thoughts that spring from "nowhere," words that come from our "intuition," and ideas that torture us relentlessly. Regardless of your beliefs, start with the most basic of all questions—Why does anything exist?

Would the world be a better place if Adolph Hitler never existed?

While many people will argue over who exactly was the worst of the worst, with names such as Pol Pot, Josef Stalin, Idi Amin, Ivan the Terrible, Genghis Khan, Nero, Osama bin Laden, Attila the Hun, and Hirohito entering the conversation, it's easy to attribute a face to the evil we all despise.

When we take a more philosophical approach and ask what the world would have looked like if our own poster child for evil had never existed, we begin seeing human progress in a whole new light. Rest assured, I'm not a fan of Hitler, or any of the other psychopaths who've splattered blood over the pages of history. But evil does play an important role, and even, sometimes, a positive one.

Our ongoing struggles with evil are never-ending, and it's up to us to stop evil wherever and whenever possible. It may be ludicrous to think we'll ever be in danger of having "too little" evil, but knowing that we are driven by adversity, and that hardship and difficulty often bring out the best in us, it's reasonable to think there may be ways to actually "optimize evil".

Our Need for Resistance

Most of us find ways to distance ourselves from all the terrible things happening in the world. Our inner voice tells us things like: "They shouldn't have lived in such a crappy neighborhood," or "If they'd raised their kid right, he wouldn't have joined that gang," or 'They should have never bought that gun."

These are all pieces of the conversations floating through our heads to convince us that we're OK and the bad stuff only happens to "those people." After all, it's easy to spot the bad guys on TV because they all look evil. Naturally everything changes when the flames of evil begins to singe the hair on our own head. Our arms-length philosophy takes on a distinctly different tone.

As humans we have an insatiable need to compete. As kids we compete for attention, we race to be first at the dinner table, and in school we compete to be smarter, better liked, better dressed, and better athletes. Once we enter the job market we compete to become more hirable, better at getting things done, and able to leap tall buildings in a single bound.

Competition brings out the best in us and pushes us farther and faster than we are ever able to motivate ourselves. Bad things also create a form of competition, forcing us to dig deeper and tap our inner resources to overcome adversity. We tend to lose our focus when there's no one to compete against, no worthy adversary. However, too much resistance is not good either. Whenever we face overwhelming odds it tends to throw us into a tailspin.

What if part of our purpose is to choose our battles, select the arenas in which we will compete to create change?

The Rise of the Cause-Architect

In 1863, Abraham Lincoln signed into law the Emancipation Proclamation, an executive order that granted some freedom to slaves.

But true freedom remained a century in the future for those living in the black vs. white world leading up to the Civil Rights movement, an effort that began in earnest in the 1950s. The movement for freeing the slaves was a social cause that tore the country apart, resulting in a civil war and a century's worth of social scarring that needed to heal before the effort could begin again.

In 1954 the stage was set with a Supreme Court ruling that made school segregation illegal. After years of marches, protests, and demonstrations, the Civil Rights movement peaked in 1963 with

Martin Luther King's famous "I Have a Dream" speech, in front of the Lincoln Memorial in Washington, DC.

After a few more tumultuous years of social unrest involving the assassinations of President Kennedy, Martin Luther King, and Bobby Kennedy, the movement came to an end in 1968 with the passage of several pieces of legislation aimed at outlawing racial discrimination.

In the past, movements like this were filled with tension and riddled with conflict. That is quickly changing. Every cause has a beginning, middle, and end. When deep-seated differences are involved, tensions will rise and fighting will occur. For most movements, where the stakes are less divisive, society simply adjusts and moves on. In the fluid society we find ourselves in today, with massive communication systems for organizing and influencing public opinion, social causes are far easier to orchestrate.

*This ease with which we can manage a movement is giving rise to a new breed of influencer—the **cause-architect**.*

As our ability to communicate, influence, and organize increases, the likelihood of violence decreases. This also means the life-cycle of most causes today will be far shorter than those of the past. More things happen quicker. It has also turned the "cause-architect," the key person serving as the movement's organizer and inspirational leader, into a respected position. Once we begin to understand the life-cycle of a cause, and the stages of activity that take place at the beginning, middle, and end, leaders can begin to manage far more organized efforts than ever in the past.

Stage One—Launching a New Cause

Whenever there are polarizing differences between two groups of people, there is an opening for a new cause to emerge. In the past, the launching points stemmed from things like poverty and wealth gaps. Today it may be caused by differences in customs, immigration standards, ethics, and values.

Those ingredients alone do not constitute a movement.

Movements begin with a single event that triggers a significant reaction, something I call a *fuse-lighting event*. This particular event will set off a chain reaction of other events leading to the creation of a stage-one social movement. For example, the Civil Rights movement grew rapidly from the reaction to Rosa Parks, a black woman, riding in the whites-only section of a bus.

Typically, social movements are created around some charismatic leader with the right combination of skills to both engineer and execute a strategy, and organize and manage a following.

After the social movement sees its first sparks of activity, there are two likely phases of recruitment. The first phase will gather the people deeply interested in the primary goal and ideal of the movement. The second wave of recruits will usually come after the movement has had some success and becomes trendy. The later recruits typically don't stick around very long.

Stage Two—Defining Success

Many movements fall apart because they are not focused around any kind of success strategy. *If there is an implicit demand for change, there needs to be a clear description of what that change will look like.* Yet many movements lack precisely this sort of vision.

The most successful movements will develop a series of benchmarks to help measure progress along the way. And they will not succeed if the whole effort is simply oriented around a need. Needs are ongoing but causes have a definable life cycle with an actual endpoint.

More important than the endpoint, definable life cycles provide definable criteria for success. Success criteria creates the foundational underpinnings of good management metrics. Tomorrow's tools will allow us to micro-analyze virtually any situation and find the primary inflection points where a change can be most effective, and good metrics can be put into place. If the metrics are measurable, progress can be tracked.

In the past, people with big hearts, who dedicate their lives to helping the needy, were held in high esteem. It was a virtuous life filled with personal fulfillment.

In the future, an even greater virtue will be bestowed upon those who are capable of solving the predicaments that create the needy class in the first place.

Cause-architects will extend their work far beyond working with the disadvantaged, and set out to wrestle social injustice to the ground. Cause architecture will become an exciting new profession well suited for inspired young people who both want to make a name for themselves and live a life of meaning.

Stage Three—Finding the End

Virtually every piece of music has a beginning, middle, and end. Much like good stories throughout history books and television scripts also have a discernible beginning, middle, and end.

Effective cause-architecture follows the same structure. The best cause- architects will be the ones who continually work themselves out of a job. Their role will be to construct a realistic action plan, execute, and complete the process of solving major social problems.

Future cause-architects will come armed with tools unimaginable by today's standards, as well as tools they invent along the way. However, the most important element in the whole equation comes with knowing when it's over. Asking for too much is as bad as asking for too little. A good cause-architect will know when they've reached the point of diminishing returns.

The Urgency of Purpose and the Forward Movement of Failure

As a futurist I spend much of my time tracking failure. Why failure? Because failures are the unforgiving anchors around which society changes directions. In the U.S. we are now witnessing a record number of failures taking place. Just look around. Failed businesses, failed systems, failed jobs, and failed marriages.

Some failures are easily predicted, where a known problem looms larger and larger until a solution is found. Most, however, are not so easy. In many respects, failures are nature's own system for checks and balances. Failures attract attention. Much like a car accident creating a gawker's block along the highway, failure attracts onlookers; some will offer to help, others will move quickly to avoid being painted with the same failure brush.

There are many forces driving the world around us, and each one of these drivers is like a hand grenade generating a blast zone of forces pushing in multiple directions. These particular forces concentrate an unusual amount of energy in the directions I've indicated here to keep this cycle in motion:

1. **Mortality Drives Urgency**
2. **Urgency Drives Purpose**
3. **Purpose Drives Our Quest for Knowledge**
4. **Our Quest for Knowledge Drives Technology**
5. **Technology Drives Complexity**
6. **Complexity Drives Failure**
7. **Failure Drives Conflict**
8. **Conflict Drives Mortality**

As we begin to study these linkages, we are able to uncover fascinating relationships, which help enormously in explaining the nature of humanity and the world we live in.

1.) Mortality Drives Urgency—The fact that we will someday die gives us only a short runway of time to get things done. The clock is ticking. We either get things done today or we lose a significant piece of the time we have left before we die. Even though people are living longer today than a hundred years ago and we have a slightly longer runway, the urgency we feel is still a prevalent force in everything we do.

While it's true that competition and our need for status also drive urgency, the constant trickle of sands falling through the hourglass leaves us feeling like our own lives are slipping through our fingers. The sound of our own mortality is a sound few can avoid listening to. Counter to what some believe, living forever may indeed be counter-productive. People who live with no end in sight may well lose their motivation for "doing anything important today."

2.) Urgency Drives Purpose—How many times have you heard someone ask, "Why am I doing this?" It's a very common concern because most of us simply despise doing anything dubbed "meaningless."

Baby-boomers are getting older. As this massive bulge in the population moves into their retirement years, many are feeling the regrets of not having lived up to their own expectations, and in doing so search for higher meaning. In what *Forbes* Magazine publisher Rich Karlgaard describes as the "Age of Meaning," the former hippie generation is now searching for a higher calling, and they want it now.

3.) Purpose Drives Our Quest for Knowledge—To find meaning and purpose, we need more knowledge. In today's world, information is infinite, but knowledge is finite. At the same time, our ability to sort through the growing storehouses of information and find those shimmering glints of needles-in-the-haystack information is a relentless quest. It is a quest we cannot do alone, and so we turn to technology.

4.) Our Quest for Knowledge Drives Technology—Human frailties and our own physical limitations drive us to find technical solutions. How can we think faster, see things outside the range of normal human vision, hear things on the other side of the world, or process information that baffles the normal mind? Virtually every invention known to mankind is an extension of human senses or human capabilities. The more information we consume, the greater our need for technology, and that's where things start getting complicated.

5.) Technology Drives Complexity—Technology drives many things, but when it comes to complexity, technology acts as the great enabler. Rather than managing a hundred accounts on paper, we can now manage a thousand accounts with a computer. Rather than spending ten hours sorting through twenty thousand books in a library, we spend ten minutes sorting through two million books online.

Technology extends our reach, but it also extends our ability to devise complex systems for managing it and complicated solutions to our problems.

Complexity itself is neither good nor bad, but it increases fragility, and too much complexity pushes us beyond our ability to manage it. And that's where things begin to fail.

6.) Complexity Drives Failure—The more complicated something is, the more likely it is to fail. Yes, in abstract terms, complexity adds function. And some measure of complexity is both necessary and beneficial. Complexity tends to function like a self-perpetuating organism. Complex systems often expand until they reach a breaking point, and that is where the conflict begins.

7.) Failure Drives Conflict—Yes, failure causes many things, but failure is very emotional, and emotional intensity leads to conflict. Our first reaction is that failure is bad, and conflict resulting from failure is even worse. Yet at the same time, failure is a time of renewal, a new branch growing where an old branch just died.

Conflict arises from our resistance to failure, and in many case we need to resist because failures are not inevitable. We only appreciate that which we struggle to achieve, and virtually every conflict clears our mind about the importance of what we are struggling for.

8.) Conflict Drives Mortality—Every conflict gives us another look into the frailties of being human. Conflicts are riddled with confusion and doubt, second-guessing and regret. They are the friction from where the rubber-meets-the-road on this turning wheel. In the end, we ask what we were fighting for, and that, in turn, drives our own feeling of mortality.

When I first sketched out this progression, I was trying to decide if it was indeed meaningful, and whether this kind of insight could be helpful. In the back of my mind I kept asking, "Is this cycle inevitable," and "Can it be stopped?" Perhaps, more importantly, "Should it be stopped?"

What I've concluded is that every step we take in life sets us up for our next step. We rarely have the insight that there is a pattern to our actions or an overarching cycle that we're part of. In most cases, we do not have the big picture ability to skip steps, circumvent them, or leap frog to a whole new section of life, only to speed them up or slow them down.

We each have many wheels to contend with. Our family wheel overlaps our business wheel, and those overlap our social and side-projects wheels. With global databases of information skyrocketing and technology improving access to it, the wheel is turning at a faster and faster pace. Every imbalance in the wheel causes a ripple effect throughout the rest of the wheel. Are we better off trying to eliminate conflict and failure, or trying to optimize it?

WITH THE NEW MANTRA BEING "FAIL FAST AND FAIL OFTEN," WE HAVE BEGUN TO ACCEPT THE INEVITABILITY OF FAILURE AS A CRUCIAL COMPONENT OF BOTH SUCCESS—AND PURPOSE.

EPIPHANY 7— OPTIMIZING LEARNING

I n the radically transformed world coming our way, learning—and the ability to continue learning—will be more essential than ever to an effective and productive life. That doesn't mean you'll be spending the rest of your life, or even a large part of it, in traditional classrooms on traditional campuses. Far from it.

In fact, many of the most highly educated people in the future may never set foot in a traditional classroom, pursuing traditional degrees, at all.

For decades, college degrees have served as a significant status symbol, one of the world's most recognizable markers for being smart. A college degree is a definable accomplishment certifying years of study; there is some validity to the notion that people who graduate from college, on average, are indeed smarter than the average non-degree holder. However, it is also clear that some of the world's most successful people took a different path and never bothered with finishing college. In fact, few people know, or care, that the sheepskin is missing from their walls. They have achieved status in other ways.

Logically then, if you are a talented person and haven't had the time, money, or opportunity to go to college, are there some legitimate substitutes for status that the rest of the world will consider to be of equal or greater value to a sheepskin?

This will be even truer—with even more non-college options and opportunities in the years and decades ahead.

Yet if you asked a cross-section of business leaders, "What other accomplishments would you consider to be as important as a college degree?" chances are they will struggle to give you an answer that doesn't have a college education somewhere in its background.

A Status Symbol Under Attack

In a recent paper, *The Future of Colleges & Universities: Blueprint for a Revolution*, I talked about how colleges were on the verge of being attacked, and one of the areas of attack is the issue of "status." College degrees remain important, but new status symbols are beginning to emerge that compete directly with the inherent status conferred by a college education. Until recently, colleges have primarily faced competition from other colleges. But even when competing, they remain unified in their evangelism for higher education.

But today, there are many status symbols able to compete with college degrees; in the future there will be many more.

What sorts of accomplishments are accessible to most people that could be construed by a potential employer, business colleague, or acquaintance as being the equivalent to a college degree, or for that matter, even better?

My answers fall into four categories:

1. **Components of Equivalency** (equal to a course or multiple courses)
2. **Equivalent to a College Degree**
3. **Better than a College Degree**
4. **Future Status Symbols**

Alternatives to going to college are definitely *not* alternatives to learning! Quite the contrary. Learning becomes an essential ingredient in virtually every path to success, but the learning process in these alternatives is far less formalized.

Components of Equivalency

Much like taking a series of courses that stack up and form the basis for a college degree, a series of smaller achievements can easily be used to form an equivalent status.

1. **Certificate Programs**—Most certificate programs are intended to either replace or supplement existing degree programs. The weight of this accomplishment varies tremendously with the institution that is granting it.
2. **Certification**—Certifications, such as Microsoft, Cisco, or Oracle Certification, have become a very popular way to bestow credentials.
3. **Apprenticeship**—The age old process of working for years under the tutelage of a master craftsman is still alive and well in some industries.
4. **Foreign Travel**—With foreign travel becoming increasingly common, it tends to hold less value today than in the past, but is still recognized as a significant achievement.

5. **Own a Patent**—Becoming a patent holder is also less rare in today's world than in the past, but is still regarded as a noteworthy accomplishment.

6. **Produce an Event**—Events range from small to huge. But a successful event, no matter the size, has the ability to position you in a way that will cause others to take notice.

7. **Memberships**—Status by association. The credibility of an association adds to the credibility of you as an individual.

8. **Start a Business**—Launching a business is a significant learning experience regardless of how successful it becomes. It also adds a new dimension to the identity of the founder.

A Degree is No Measure of Who **You** Are— or of All You've Learned

As part of a family, your domestic life and the relatives closest to you measure who you are as an individual. As a prospective employee, you are evaluated by your skills, talents, and knowledge. As part of a community, you are gauged by the kind of relationships you build and maintain. As an athlete you are assessed by your physical strengths, your reaction times, and your determination. Whatever kind of lens or filter we place over our lives, we use different systems for measuring those key differentiators. And while we all think we are the world's foremost expert on ourselves, we actually know very little.

That's about to change. The Internet of Things is already comprised of over ten billion moving parts, and by 2020 that number will grow to over fifty billion.

These "things" have a way of gathering information about ourselves in ways we never imagined were possible. Not only will we be able to monitor the quantity and quality of food we eat, the air we breathe, and our daily activities, but we will also be tracking the information we consume, our moods, our level of engagement, and what undertakings we find most stimulating.

In addition to charting the normal inputs and outputs for our mind and body, we will also be evaluating the context in which we exist. Whether it's an emotional context, environmental context, or spiritual context, each plays an important role in determining who we are. In the future, it all becomes measurable. The "quantified self" is all about building a vast and measurable information sphere around us. As we get better acquainted with the Delphic maxim "know thyself," we will become far more aware of our deficiencies and the pieces of learning needed to shore up our shortfalls. That's why this will have such a tremendous impact on colleges.

Compensating for these deficiencies won't be about getting Bachelor or Master's degrees. Rather, they will be about gaining experiences, reading books, meeting people, or working as an apprentice. Some of us will take courses at a university, but not an entire degree package.

In other words, what you've learned and what you know will be far more important than what you've *studied*.

Quantifying Human Attributes

If you were performing a job search for someone who is extremely creative, detail oriented, or has a great passion for life, what kind of credentials will you be looking for?

If you need someone who is extremely persistent, enjoys working in isolation, or an ability to discern tiny little details, what kind of diplomas will you want them to have?

Human attributes fall into many different categories that, when connected together, form a nearly infinite number of combinations. Yet, it is these same highly nuanced human characteristics that are often leveraged to our advantage, that also become a huge liability in a system that can't quantify them.

Not only can we not measure, rate, or score human attributes, we currently have no well-accepted system for improving on them or credentialing them.

The big picture of who we *are* gets lost in a blur of anecdotes about what we have *done*. Emerging from the shadows of yesteryear's murkiness comes a host of new quantifiable-self technologies that promise an end to the primary-colors-only view of our uniqueness.

Imagine stepping through a series of assessments that rates you in, say, 947 different categories of physical attributes and human characteristics.

Once you have your personal information sphere in place, now visualize a similar but larger sphere that depicts your goals and desires for the future and plots out a tactical plan for getting there.

As an example, if you felt you wanted more control over your life, it might recommend a series of management books, videos, or classes to help you gain those skills. If you have a secret desire to become more famous, it might recommend a number of achievable benchmarks that would position you in the limelight. If your interests center on becoming more physically trim and active, it may advise you on possible workout regiments, diets, and workshops.

Every person's quantifiable information sphere would be superimposed on their desired goal-sphere, and system algorithms would constantly be prompting you on ways to get closer to your goal.

In this real-time quantifiable-self machine, every time your interests, desires, or ambitions change, so will your goal-sphere. In fact, it will recalculate many times a second to reflect the dynamic nature of matching personality shifts with new gap-filling options.

As the quantifiable-self catches on, the tools for human assessment will expand exponentially. As employers lose confidence in traditional transcripts and college degrees as a predictor of success, they will turn towards more sophisticated attribute-matching systems for sorting through the ultra-granular quantifiable-self and finding the closest fit. People who don't make the shortlist for a job opening will be given an auto-generated overview of their perceived deficiencies and ways to improve upon them.

The reason this will have such a profound effect on colleges is because our credentialing systems today for granting credits and degrees will have virtually no standing in the hyper-granular metrics used to measure job candidates in the future.

Looking at this through a bigger picture lens, as we move into the quantifiable self-era, we will have far more tools for taking charge of our own destiny. Rather than spending much of our future income on a path of discovery by taking marginally relevant courses at a university, we will have a wide array of super-tools for both finding and engaging in personally relevant experiences, all in an effort to optimize "you" to become the "ultimate you."

One key super-tool for learning may be the Micro-College.

Programming a Micro-College

In 2012 the DaVinci Institute launched a computer programmer training school, DaVinci Coders. DaVinci Coders is an 11-13 week, beginner-based training in Ruby on Rails, patterned after the successful Chicago-based school, Code Academy (later renamed The Starter League).

One of the key people we tapped to join our world-class team of instructors was Jason Noble.

Working as a Senior Software Engineer for Comverge, an intelligent energy management company in Denver, and also part-time instructor for DaVinci Coders, Jason understands what it takes to train people both in the classroom and on the job. He has compared the apprenticeship times necessary to bring three different newly hired Junior Developers up to speed—one with no Rails experience, one who attended our

11-week course, and another who attended a 26-week program at a different school.

Jason concluded that the one with no Rails experience required 6-7 month apprenticeship time, the one with 11-weeks training required 2 months, and the one with 26 weeks schooling was up to speed in 3 weeks. He also estimated hiring a talented college grad with a computer science major, the apprenticeship time would likely be more than 2 months, but they'd also bring other valuable tools to the table.

Admittedly, this is an unusually tiny sample size for a test case, and training times will vary greatly. But this type of comparative analysis naturally begs the question of how much training should be required prior to taking a job, as well as whether the investment of time and money spent on training should be optimized around the company or the employee, acknowledging that there will always be some in-house training required.

When we look at the bigger picture of retraining for this, and many other professions, knowing that people will be rebooting their careers far more often in the future, with time being such a precious commodity, how do we create the leanest possible educational model for jobs in the future?

That's where the Micro-College concept comes into play.

How Lean is Too Lean, and How Fat is Too Fat?

In the movie "*Slumdog Millionaire*," Jamal Malik, a penniless eighteen-year-old orphan from the slums of Mumbai, is asked a series of very difficult questions on his quest to win a staggering twenty million-rupee prize on India's *Who Wants to Be a Millionaire?* As a street-smart kid with virtually no formal education, the probability of him answering these questions correctly was zero. However, as luck would have it, his life experience had given him precisely the answers he needed, if very little more.

This is an example of an extremely narrow education, only applicable to the one-in-a-billion situation portrayed in the movie. On the other side of the equation are people who go through all the work of getting Bachelor's and Master's degrees, yet and still don't possess the skills necessary to gain employment. Traditional colleges, for the most part, do a great job with courses primarily oriented around seat time. Traditional colleges also embrace the overarching philosophy that nothing of value can be learned in less than four years, a timeframe woefully out of sync with someone needing to change career paths in today's frenetic environment.

At what point is education "too lean," and conversely, when is it "too fat?"

Typically, universities require students to achieve both breadth of knowledge across disciplines, and depth of knowledge in a chosen subject area, their major. For this reason, students studying Arts or Humanities are required to take science courses, and vice-versa. While this made sense hundreds of years ago when the university system was first created, and the body of knowledge was far smaller, the average person today in the U.S. spends 11.8 hours each day consuming information. Much of this is TV, radio, and other frivolous varieties of information, but not all of it.

> "Does 'breadth of learning' still need to be
> a requirement since breadth is already part
> of our 'ambient learning culture?'"

The sheer volume of information we're exposed to every day makes the average person today far more informed, aware, and intelligent than their counterpart 50 years ago. Known as the "Flynn Effect," after researcher James R. Flynn, the average IQ in the U.S. has been increasing every generation for over eighty years, ever since IQ tests were first developed.

Possible Types of Micro-Colleges

Micro-Colleges are any form of concentrated post-secondary education oriented around the minimum entry point into a particular profession. With literally millions of people needing to shift careers every year, and the long, drawn out cycles of traditional colleges being a poor solution for time-crunched rank-and-file displaced workers, we are seeing a massive new opportunity arising for short-term, pre-apprenticeship training. Many Micro-Colleges will fall into the category we often refer to as vocational training, a term poorly suited for the professional craftsmen, artisans, and technicians they will be producing. Since status and credentialing are critical elements of every career choice, any training producing specialized experts will need to come with industry-recognized certifications and titles.

On a "Future of Beer Tour," an event we produced at the DaVinci Institute that took us on a futuristic bus tour of 5 local craft breweries, one of our on-board experts mentioned that a local college was planning to offer an official major and degree for becoming a "brewmaster." This is yet one more example of taking an industry where most brewmasters are self-taught in a couple months and stretching it into an expensive 4-year college degree. The Micro-College approach to training brewmasters would be an intense 2-4 month training program with a designated apprenticeship period learning on the job.

Using this line of thinking, the potential for Micro-Colleges is huge, and emerging technologies and business trends are creating more opportunities on a regular basis. Here are a few possibilities:

- **Certified crowdfunding training**
- **Dog breeder university**

- **Brewmaster college**
- **3D print technician training center**
- **Drone pilot school**
- **Body scanner academy**
- **Data visualization and analytics school**
- **Aquaponics farmers institute**
- **Online competition manager/producer school**
- **Project manager training for the freelance economy**
- **Urban agriculture academy**
- **School for legacy management consultants**
- **Pet day care management school**
- **3D food printer chef institute**
- **Privacy management academy**
- **Senior living management school**

Creative team brainstorming could easily come up with more than 100 possible Micro-Colleges—in the near future that number may reach well into the thousands!

The "Engineering Major" Scenario

As a former IBM engineer, I've thought a lot about the relevance of my college years and the work I did as an engineer. Since my coursework happened in the pre-computer era, most of the skills I needed after computers were introduced were primarily self-taught. While I used a fair amount of math, trig, and geometry, I never had to call upon or apply what I learned in the required higher-level math courses like calculus and differential equations.

Most of my engineering coursework quickly became dated as computers and calculators made slide rules, protractors, calipers, and drafting tables obsolete. My first FORTRAN class using a card-punch machine was obsolete even before I punched my last card. Perhaps the most valuable courses with long-term relevance taught writing, English, speech, art, design, as well as the special research projects that forced me to find my own answers and write a final report presenting my findings. Art classes helped me understand that engineering was a form of creative expression. Nothing I learned was worthless, but certainly some courses held far greater value than others, and it all boiled down to opportunity cost. Is this course worth what I'm paying for it, and could I be learning something more valuable elsewhere?

Starting from the premise of training for the minimum skill requirements of a profession, we should determine the precise core courses needed for someone to enter a particular field, engineering, for example. An electrical engineer is far different from a petroleum engineer, and a mechanical engineer differs again. So the number and type of core courses may vary. But rather than expanding courses to fill an arbitrary four-year requirement, we need to decide how much fat can be trimmed from the curriculum while still producing an effective, competent engineer?

Using an intensive, full-immersion approach to education, could a school churn out competent industry-ready engineers in less than 2 years? If the school were tied to an industry-specific apprenticeship program with a near-perfect handoff between academia and real-world work happening inside the industry, what would a super-lean engineering program like this look like?

The Coming Transition

The systems used to create colleges centuries ago seem justifiably primitive by today's standards. Learning formulas for nearly every degree are based on hours spent in class, one of the least important considerations when it comes to assessing talent.

Colleges today cost far too much, and they take far too long. And now, like many other industries, traditional colleges are being tasked to do more with less. But few colleges today have anything like a clear understanding of what "less" looks like. MOOCs are offering a new way to produce and distribute lecture-style courses, but that only represents a piece of a much larger equation.

Because of Micro-Colleges' ability to position themselves instantly at the critical cross-section of skill and commerce, far more new industries will be born through Micro-Colleges than through traditional colleges.

Since we launched DaVinci Coders in the second quarter of 2012, over 250 other coder schools have cropped around the U.S. and even more in Canada and Europe. Every successful Micro-College prompts others to follow in their footsteps and refine the original business model. It's easy to imagine that as traditional colleges see their student base decline, many will begin to partner, merge, and purchase fledgling Micro-Colleges and begin incorporating these new approaches to study into their own catalog of course offerings.

Since existing colleges bring with them credit-granting accreditation, along with status, credibility, and the ability to offer student loans, in-house Micro-Colleges will likely become a rapidly growing part of campus life. Many colleges will find the Micro-College niche they take on to be the key differentiator between them and other schools. Using the school-within-a-school approach, core Micro-College programs will become feeder mechanisms for additional types of credentialing.

Will there be a Micro-College in your future?

Preparing Our Minds for Thoughts Unthinkable: The Future of Colleges and Universities

If you haven't noticed, there's a massive battle brewing in academia. It's not just a battle between next generation and traditional education. What's at stake is nothing short of the future of humanity. Next generation academic systems will determine, from here on out, how the human brain gets developed. The stakes have never been higher. Our descendants are counting on us.

So far, our best and brightest have fallen short. We have not been able to cure cancer, prevent natural disasters, or stop corruption. The challenges ahead will be even greater. As we plan for the future, we need to set our sights on producing a caliber of people who are exponentially better than we are. And we do that by creating innovative new systems to take us there.

It's OK to mourn the loss of our old academic institutions, but while mourning, we need to admit they've been holding us back. Traditional education has kept us tethered to our comparatively small potential, minor accomplishments, and tiny victories as we lose sight of what's truly important in the clouds of minutia and complexity surrounding us today.

As a backward-looking society, we wish to emulate the heroes of the past, using their achievements as the symbolic gold standard for us to live up to. However, the standard-bearer for significance in the future will be a thousand fold greater. Our backward-looking obsession with problems will all but disappear with forward-looking accomplishments in the future.

I'm hoping to set the stage for what I believe to be humankind's most important opportunity, the opportunity to build a better grade of human.

The 10X Speed-Learning Scenario

Consider the following scenario. In 2020 a system is invented for amping up learning speeds by a factor of ten. Any person who spends just one hour a day with this learning system can learn the equivalent of a Bachelor's Degree's worth of knowledge in less than a year without having to compromise their lifestyle.

Starting at age ten, committing one hour a day to learning, this individual will earn the equivalent of forty-four bachelor's degrees by the time he or she turns eighty. In the U.S. the average tuition cost of a Bachelor's Degree at a public four-year college is $102,352. Multiplied by forty-four degrees at today's rate, the cost would be over $4.5 million. However, in this scenario the cost of learning is reduced to $10 per one-hour course, or $3,650 per year. Over the course of a lifetime, total learning costs would be $255,500.

Learning at this speed, the person would consume the equivalent of eighty-one three-credit courses each year, or 5,677 three-credit courses over their lifetime.

MIT, as an example, offers roughly two thousand different courses. This person would be consuming close to three times as many courses as MIT currently offers.

If you think this scenario is out of line, consider that virtually every industry throughout the world is continually being forced to do more with less. A 10X improvement in computer speeds, agriculture efficiencies, or steel production happens every few years. For those who think it's not possible to achieve this kind of efficiency in education, the answer lies in devising new systems, ones without the current self-imposed limits.

Redistributing Educational Wealth

Creating and distributing online courses is not easy. Course creation needs to recognize the creative genius behind the process and somehow remunerate those who make it possible.

A few years ago I wrote about the concept of "fractal transactions," where financial transactions are automatically subdivided to automatically pay for a variety of contributors to a particular product or service. The advantage of this arrangement is that it eliminates all money going to one person or company, who then has to pay lower level helpers in a timely manner.

When this approach is used to pay for courseware, the revenue stream generated by each purchase will be divided among the courseware creator, distribution company, online courseware builder, official record archive, and much more. Courseware prices need to be kept low to make courseware accessible to anyone interested in learning.

If we use a $10 purchase price for a course, here is an example of how funds could automatically be distributed to contributing entities:

40%—Courseware Creator ($4.00)

25%—Promotion and Distribution Company ($2.50)

10%—Online Courseware Builder ($1.00)

5%—Official Record Archive ($.50)

3%—Smart Profiler ($.30)

3%—Multidimensional Tagging Engine ($.30)

3%—Recommendation Engine ($.30)

3%—Learning Methodology ($.30)

5%—Financial Transaction Costs ($.50)

3%—Future Contributors ($.30)

NOTE: If the $10 purchase price for a one-hour course seems high, keep in mind that accelerated learning processes will make this equivalent to ten hours' worth of classes today.

Currently distribution companies like Coursera, EDx, and Udacity occupy a controlling position. If they choose to implement a participative wealth strategy like this, the strategy can quickly become the de facto standard across the industry. Admittedly, I've greatly oversimplified the situation and overlooked the value of many contributors, but this was intended simply as a starting point. Once a system like this is introduced, even tiny fractions will empower entire industries to do better.

Rapid Courseware Creation

I have predicted that 50% of colleges will collapse by 2030, and the fallout from these failures will not be pleasant. However, as with all predictions, the fate facing these institutions is not inevitable. A few will find a way to navigate through the radical transformation ahead, but it helps to have a clearer picture of what tomorrow will bring.

Colleges now have a far greater calling than simply delivering courses. Professors lecturing in the classroom will still exist for many years to come, at least at those schools that remain in business, but the resources of academia are far too valuable to waste on repetitiously presenting the same class over and over again.

One of their most valuable skills, that doesn't get used nearly enough, is the ability to create new courseware.

Similar to the way television networks unveil their "new fall lineup," next generation colleges will periodically unveil a new courseware series.

Here are a series of six scenarios to better explain how courseware development teams will work:

Scenario #1: A rapid courseware developer's package will be created that enables colleges to create their own courses and make money from every sale. Each course is framed around a standardized 60-minute format with a variety of media inputs. Courses can be tagged with approvals by institutions, rated by students, and framed around a personalized adaptive learning engine.

Scenario #2: Courseware rating systems will be developed to add integrity to the rapidly evolving system. Rating systems will be structured as a checks-and-balance system where individual groups, colonies, or rating services can create their own authority and place tags of approval or disapproval on courses. These tags will be a central feature of the search criteria used by smart student profilers and courseware recommendation engines. As example, a person may only want to take courses approved by an association like IEEE, a particular university, a church body, or political group.

Scenario #3: Colleges that focus on research will have a built-in advantage and will leverage each research project by spinning off a series of new courses surrounding the research. Projects will not only develop their own revenue stream through new courses, but the research will also attract many new students. Tech transfer efforts will be aided by the courseware as well. Courseware will become a broadcast medium through which others will learn about new technologies as well as related opportunities. With the added publicity from effective new courses, government and corporate grants will become more readily available to fund research.

Scenario #4: Colleges will aggressively seek out research projects to better inform us of the world to come. Whenever a natural disaster strikes, news teams serve as the first wave of information about what just happened. In the future, college research teams will serve as the second wave. Every famine, hurricane, plane crash, and tidal wave will attract several college teams, each looking at different aspects of the situation. With research teams scouring the earth for new projects, some will look at a tidal wave from a physics-hard sciences approach, such as the precursors to wave formation, others will look at the economic impact, political turmoil, social shifts, and other long-term generational effects it may have on a community's customs, language, and neighboring influences.

Scenario #5: Colleges will begin to orient their business around lifelong relationships with their students. Some traditional courses will still exist and others will be oriented around short on-campus experiences such as two-week learning camps. But a growing portion of the learning will take place online. Since the number of people setting foot on campus will be dropping, successful colleges will begin adding more research and experiential learning components to their offerings.

Scenario #6: Colleges will begin experimenting with higher and higher achievement levels. In recognition of learning that will take place over a lifetime, degrees and diplomas will be created for extreme and super extreme levels of learning. Master's and PhDs will only be junior-level accomplishments on these new rating scales. With learning made easy and expanded over a lifetime, colleges will be able to capitalize on new ways for individuals to differentiate themselves from the masses. Diplomas will become as individualized as the accomplishments they reflect. These uber-diplomas will become an ongoing driver for continued involvement and serve as enduring revenue streams for the institution.

These six scenarios are intended as a tool for gaining a new perspective on what may be possible. In the future the largest web property on the Internet will be oriented around education.

As the price of education drops, people will begin to "consume" far more education. In our increasingly competitive work environments, with people from around the world competing for the same work we do, adding new skills to a future credentialing system will become an everyday occurrence.

In the U.S. we have a total of 4,495 degree-granting institutions sharing the tuition money being paid by over twenty million students annually. It is a growing system built on easy money, and possessing great inertia. But public higher education is changing, and it's changing in some very fundamental ways whether we like it or not. The forces driving these changes aren't simply financial; they reflect major shifts in student attitudes, expectations, and demands.

For those associated with an Ivy League College where the acceptance rate for incoming freshmen is under 8%, it's easy to ignore forecasts of change. But while the Ivy Leaguers may be safer, many smaller institutions are on a collision course with destiny. The students of tomorrow will need to prepare themselves for a higher calling. This higher calling will be to pre-empt crises before they occur, anticipate disasters before they happen, and solve some of mankind's greatest problems, starting with the problem of our own ignorance and shortcomings.

Much like a person walking through a dark forest with a flashlight that illuminates only a short distance ahead, each step forward gives us a new perspective by extending the light into areas previously dark. The students of tomorrow will be our bigger flashlights. Until now, ours has been a dance with the ordinary.

> "History shows us that we are immersed in cycles, systems, and patterns that repeat again and again. Tomorrow's history books will show us that all patterns are made to be broken, and all cycles waiting to be transformed."

Colleges will need to position themselves on the bleeding edge of what comes next. We will always need the backward-looking perspectives in order to understand where we have come from, but a new breed of visionaries, bestowed with unusual tools for preempting *future* disasters, will become our most esteemed professionals.

Future colleges and their equivalents will become our checks and balance for the status quo.

The grand mission for colleges, and for learning in general, in the future may well be phased as:

PREPARING HUMANITY FOR WORLDS UNKNOWN, PREPARING OUR MINDS FOR THOUGHTS UNTHINKABLE, AND PREPARING OUR RESOLVE FOR STRUGGLES UNIMAGINABLE.

EPIHANY Z— OPTIMIZING OUR TOOLS

As humans, we've become very adept at using tools. We've used them to build the world we live in today—and even more to build the world of tomorrow.

But one of the things we've always done with our tools is use them to build better tools.

What happens when the tools themselves are able to build their own better versions? What happens to *us* when the better versions of our tools begin to take our jobs?

Hi, I'm a Robot, and I'm Here to Take Your Job

In September 1989, GE Chairman Jack Welch flew to Bangalore, India, for a breakfast meeting with an Indian delegation that included Prime Minister, Rajiv Gandhi. The purpose of his trip was to sell airplane engines and medical equipment to India, but the meeting took an interesting twist along the way.

Rather than buying what GE had to sell, the Prime Minister Gandhi proposed that GE buy software from India. After looking at the amazingly low labor costs, Welch decided instead to outsource portions of its business, starting with Bangalore's first call center. This short meeting led to an outsourcing revolution that would dramatically transform both the Indian and U.S. economies.

We are now on the verge of another business transformation, but this time workers are not being replaced by low cost labor in other countries. Rather, machines are replacing them.

Science fiction writers have led us to believe that humanoid robots, with all the nuanced skills and talents of humans, would be walking among us today. But rather than some Stepford Wife-like creations appearing at our doors and telling us they were taking our jobs, the true job-stealing culprits have been far sublter, appearing under the guise of automation, without establishing any clear relationship between the machines and the people they're replacing.

Hidden inside this menacing movement to displace labor is a far more complicated shifting of social order. What appears on the outside to be little more than executives with blinders chasing higher profits may instead be humankind's biggest opportunity.

The Displacement Myth

One common fallacy is that machines are replacing people. The reality is machines don't work without humans. A more accurate description is that a large number of people are being replaced by a smaller number of people using machines.

Automated machines, robots, and other devices are designed to make people more efficient, but there is never a 100% replacement ratio.

Driverless cars, as an example, will eliminate the need for human drivers, but will still require skilled maintenance and repair technicians, operations managers, logistics people for dealing with failing vehicles, customer service representatives, etc.

Pilotless planes will still need ground crews, station chiefs, maintenance crews, and more.

Teacher-less schools will still need course designers, on-site coaches, software teams in the background, and much more.

Even workerless businesses will still require owners and support staff to direct the efforts of the business.

While it may be conceivable that the human replacement ratio could, on occasion, be dramatic, pushed as high as 1,000 to 1. But most of the time it will be far less. At the same time, a super-efficient society will have the ability to accomplish far more than ever in the past.

Moving into an Era of Super-Efficient Humans

Today's workers are being replaced by far more efficient workers who are capable of leveraging machines and other forms of automation. Rather than having someone show up with a magical machine under their arm that can do everything you currently do, the machines I'm referring to are a combination of computers, software, communication networks, automated devices, mobile apps, and the Internet. Perhaps there's even a robot or two thrown into the mix. Low-skilled workers of the past are being replaced by those capable of operating a myriad of software and devices, born with the tech instincts to master whatever new machinery, system, or technology gets thrown into the mix.

The bottom line is the work being done today will require far fewer workers in the future and today's workers will be reskilled for work in new industries.

The Work of the Future

Rest assured, there will always be more problems than we have solutions for. Since virtually every solution generates additional problems, the area of problem-solving alone has a seemingly infinite number of opportunities that lie ahead.

In addition to fixing our current ailments, many will opt instead to pursue a higher calling, and these will include a myriad of possibilities:

1. **Cures**—In the medical world we need to step past treating the ailments and focus on long-term cures. These include cures for cancer, AIDs, MS, epilepsy, heart disease, stroke, diabetes, Alzheimer's, dementia, and many more. Some will even focus on ending human aging altogether, an area with strong near-term potential.

2. **Natural Disasters**—We have an obligation to somehow mitigate the impact of natural disasters. This will include efforts to stop forest fires, hurricanes, earthquakes, avalanches, tornadoes, hail, and flooding—to name just a few.

3. **Correcting Deviant Behavior**—Many among us experience-traumatizing events that cause personalities to skew far from society's norm. Others have brain defects that cause outrageous behavior. To some, these are the problems most deserving of their time and attention.

4. **Colonizing Other Planets**—Many believe the human race cannot survive if all humans only live on one planet. Traveling to distant worlds has been the lifelong dream of many and living in a super-efficient society will bring that dream ever closer to reality.

5. **Ending Extreme Poverty**—Too much of humanity is still slipping between the cracks. A fully engaged world puts everyone to work, not just the gifted few.

6. **Discovery and Exploration**—Even with all our scientific advancements, we still don't know what's inside the earth or what gravity is. At the same time we are discovering new species of fish, animals, insects, and birds on a regular basis. When it comes to discovery and exploration, we've only scratched the surface.

7. **Trailblazing Firsts**—Few of us remember the second person to set foot on the moon, or the second person to invent the airplane, or the second one to run a mile in under four minutes. We place a disproportionate amount of attention on those who go first, and there are a lot of "firsts" that still need to be accomplished.

8. **Extending Human Abilities and Capabilities**—Human awareness ends at the outer reaches of our capabilities. We have little understanding of distant universes, sub-atomic particles, and other dimensions. Extending human abilities and capabilities will open doors in places we didn't know doors existed.

101 Endangered Jobs by 2030

Business owners today are actively deciding whether their next hire should be a person or a machine. After all, machines can work in the dark and don't come with decades of HR case law requiring time off for holidays, personal illness, excessive overtime, chronic stress or anxiety.

If you've not heard the phrase "technological unemployment," brace yourself; you'll be hearing it a lot over the coming years. Technology is automating jobs out of existence at a record clip, and it's only getting started.

My predictions of endangered jobs will likely strike fear into the hearts of countless millions trying to find meaningful work. But while crystal balls everywhere are showing massive changes on the horizon, it's not all-negative news.

For those well-attuned to the top three skills needed for the future—adaptability, flexibility, and resourcefulness—there will be more opportunities than they can possibly imagine. This is precisely the shift in perspective we're about to go through as the tools at our disposal begin to increase our capabilities exponentially.

As I describe the following endangered jobs, understand there will be thousands of derivative career paths ready to surface from the shadows. We live in an unbelievably exciting time, and those who master the fine art of controlling their own destiny will rise to the inspiring new lifestyle category of "rogue commanders of the known universe."

Cause of Disruption: Driverless Cars

When DARPA launched their first Grand Challenge in 2004, the idea of autonomous driverless vehicles for everyone seemed like a plot for a bad science fiction novel about the far distant future. The results of the first competition bore that out with few of the entrants getting past the starting blocks.

Between now and 2030, driverless features will pave the way for fully autonomous vehicles and the demand for drivers will begin to plummet. On-demand transportation services, where people can hail a driverless vehicle at any time will become a staple of everyday metro living.

Endangered Jobs
  <u>Drivers</u>
  1. **Taxi Driver**
  2. **Limo driver**
  3. **Bus drivers**
  4. **Rental car personnel**
  <u>Delivery Positions</u>
  5. **Truck drivers**
  6. **Mail carriers**

Public Safety
7.   Traffic cops
8.   Meter maids
9.   Traffic court judges
10.   Traffic court lawyers
11.   Traffic court Das
12.   Traffic court support staff misc.
13.   Parking lot attendants
14.   Valet attendants
15.   Car wash workers

Cause of Disruption: Flying Drones

Flying drones will be configured into thousands of different forms, shapes, and sizes. They can be low flying, high flying, tiny or huge, silent or noisy, super-visible or totally invisible, your best friend, or your worst enemy. Without the proper protections, drones can be dangerous. The same drones that deliver food and water can also deliver bombs and poison. We may very well have drones watching the workers who watch the drones, and even that may not be enough.

Endangered Jobs
Delivery Positions
16.   Courier service
17.   Restaurant delivery
18.   Grocery delivery
19.   Postal delivery
Agriculture
20.   Crop monitors/consultants
21.   Spraying services
22.   Shepherds
23.   Wranglers/herders

24. **Varmint exterminators**

Surveying

25. **Land and field surveyors**
26. **Environmental engineers**
27. **Geologists**
28. **Emergency Rescue**

Emergency Response Teams

29. **Search and rescue teams**
30. **Firefighters**

News Services

31. **Mobile news trucks**
32. **Construction site monitors**
33. **Building inspectors**
34. **Security guards**
35. **Parole officers**

Cause of Disruption: 3D Printers

3D printing, often described to as additive manufacturing, is a process for making three dimensional parts and objects from a digital model. 3D printing uses "additive processes," to create an object by adding layer upon layer of material until it's complete. Manufacturing in the past relied on subtractive processes where blocks of metal, wood, or other material is removed with drills, laser cutters, and other machines until the final part was complete. This involved skilled machine operators and material handlers.

3D printing reduces the need for skilled operators as well as the need for expensive machines. As a result, parts can be manufactured locally for less money than even the cheapest labor in foreign manufacturing plants.

This technology is already being used in many fields: jewelry, footwear, industrial design, architecture, engineering and construction, automotive, aerospace, dental and medical industries, education, geographic information systems, civil engineering, and many others.

Endangered Jobs
Manufacturing
36. **Plastic press operators**
37. **Machinists**
38. **Shipping & receiving**
39. **Union representatives**
40. **Warehouse workers**

Cause of Disruption: Contour Crafting
Contour Crafting is a form of 3D printing that uses robotic arms and nozzles to squeeze out layers of concrete or other materials, moving back and forth over a set path in order to fabricate large objects, such as houses. It is a construction technology that has great potential for low-cost, customized buildings that are quicker to make, reducing energy and emissions along the way. This type of technology will have major implications for all construction, building, and home repair jobs.

Endangered Jobs
Home Construction
41. **Carpenters**
42. **Concrete workers**
43. **Home remodeling**
44. **City planners**
45. **Homeowner insurance agents**
46. **Real estate agents**

Cause of Disruption: Big Data and Artificial Intelligence

There is an increasingly blurred line between big data and artificial intelligence. One of the dangers of artificial intelligence (AI), is that it has the potential for outsmarting humans in the financial markets. Elon Musk made headlines when he said artificial intelligence could be "unleashing the demons," and researchers from some of the top U.S. universities say he's not wrong. In spite of growing fears, AI will be entering our lives in many different ways, ranging from smart devices, to automated decision-makers, to synthetic designers.

Endangered Jobs

<u>Writing</u>

47. **News reporters**
48. **Sports reporters**
49. **Wall street reporters**
50. **Journalists**
51. **Authors**

<u>Military</u>

52. **Military planners**
53. **Cryptographers**

<u>Medical</u>

54. **Dietitians**
55. **Nutritionists**
56. **Doctors**
57. **Sonographers**
58. **Phlebotomists**
59. **Radiologists**
60. **Psychotherapists**
61. **Counselors/psychologists**

Financial Services
62. **Financial planners/advisors**
63. **Accountants**
64. **Tax advisors**
65. **Auditors**
66. **Bookkeepers**

Legal Services
67. **Lawyers**
68. **Compliance officers/workers**
69. **Bill collectors**

Miscellaneous
70. **Meeting/event planners**
71. **Cost estimators**
72. **Fitness coaches**
73. **Logisticians**
74. **Interpreters/translators**
75. **Customer service reps**
76. **Teachers**

Cause of Disruption: Mass Energy Storage

Any form of mass energy storage will dramatically improve renewable energy's role in the marketplace. The first companies to commercialize utility-scale energy storage stand to make a fortune and pioneer some of the most significant advancements to the world's power generation and distribution system in decades. While we are not quite there yet, significant technological breakthroughs are on the horizon and major energy storage installations will soon become commonplace.

Large-scale methods of storing energy include flywheels, compressed air energy storage, hydrogen storage, thermal energy storage, and power to gas. Smaller scale commercial application-specific storage methods include flywheels, capacitors and super capacitors. In five to ten years the mass, grid-scale, bulk energy storage industry will likely be a rapidly growing industry, much as solar and wind are today. Electricity generated but not consumed is a waste of natural resources and money lost. Energy storage will change all that.

Endangered Jobs

77.   **Energy planners**
78.   **Environmental designers**
79.   **Energy auditors**
80.   **Power plant operators**
81.   **Miners**
82.   **Oil well drillers, roughnecks**
83.   **Geologists**
84.   **Meter readers**
85.   **Gas/propane delivery**

Cause of Disruption: Robots

Robots taking jobs from manufacturing workers has been happening for decades. Rapidly advancing software will spread the threat of job-killing automation to nearly every occupation.

Anything that can be automated will be.

A robotic "doc-in-a-box" will help diagnose routine medical problems in many areas, while other machines will perform surgeries and other procedures. If the human touch is not essential to the task, it's fair to assume that it will be automated away. Over the coming decades, robots will enter the lives of every person on earth on far more levels than we ever dreamed possible.

Endangered Jobs

Retail

86. **Retail clerks**
87. **Checkout clerks**
88. **Stockers**
89. **Inventory controllers**
90. **Sign spinners**

Medical

91. **Surgeons**
92. **Home healthcare**
93. **Pharmacists**
94. **Veterinarians**

Maintenance

95. **Painters**
96. **Janitors**
97. **Landscapers**
98. **Pool cleaners**
99. **Grounds keepers**
100. **Exterminators**
101. **Lumberjacks**

The question remains: Will technology become a net-destroyer of jobs or a net-creator? For each of the endangered jobs listed above, I can easily come up with several logical offshoots that may amount to a net increase in jobs.

As an example, traditional lawyers may transition into super-lawyers handling ten times the caseload of lawyers today. Limo drivers may become fleet operators managing fifty to one hundred cars at a time. Painters may become conductors of paint symphonies with robot painters completing entire houses in less than an hour.

If it costs a tenth as much to paint your house, you'll simply do it more often. This same line of thinking applies to washing your car, traveling around the world, and buying designer clothes.

As today's significant accomplishments become more common, mega-accomplishments will take their place. We need to set our sights on far more of tomorrow's "mega-accomplishments." It is simply not possible to run out of work to do in the world. But, whether or not there will be a job tied to the work that needs to be done is another matter entirely.

The Growing Dangers of Technological Unemployment and the Re-skilling of America

When Facebook announced the $2 billion acquisition of virtual reality innovator Oculus Rift, they not only put a giant stamp of approval on the technology, but they also triggered an instant demand for virtual reality designers, developers, and engineers.

The same was true when Google and Facebook each announced the acquisition of a solar powered drone company, Titan and Ascenta respectively. Suddenly we began seeing a dramatic uptick in the need for solar-drone engineers, drone-pilots, air rights lobbyists, global network planners, analysts, engineers, and logisticians.

Bold companies making moves like this are instantly triggering the need for talented people with skills aligned to grow with these cutting edge industries.

Whether it's Tesla Motors announcing the creation of a fully automated battery factory, Intel buying the wearable tech company Basic Science, Apple buying Dr. Dre's Beats Electronics, or Google's purchase of Dropcam, Nest, and Skybox, the business world is forecasting the need for radically different skills than colleges and universities are preparing students for.

In these types of industries, it's no longer possible to project the talent needs of business and industry 5-6 years in advance, the time it takes most universities to develop a new degree program and graduate their first class. Instead, these new skill-shifts come wrapped in a very short lead-time, *often as little as 3-4 months.*

Udacity's founder, Sebastian Thrun, announced his solution, the NanoDegree, where short-course training is carefully aligned with hiring companies, and virtually everyone graduating within the initial demand period is guaranteed a job. Udacity's NanoDegrees are very similar to the Micro College programs being developed by the DaVinci Institute that can rapidly respond to swings in the corporate training marketplace.

Here's why NanoDegrees and Micro Colleges are about to become the hottest of all the hot topics for career-shifting people everywhere.

The Growing Dangers of Technological Unemployment

Peter Diamandis, founder of Singularity University and the X-Prize Foundation, invited me to a 2-day summit along with some of Silicon Valley's best thinkers to discuss future jobs and the growing dangers of technological unemployment.

In Peter's way of thinking, even though we are headed toward a world of abundance, having a significant loss of jobs due to robots and automation has the potential of causing a near term backlash.

Every time 10,000 people are laid off their jobs, it creates a glass-half-full-half-empty kind of dilemma. The layoffs increase our pool of available human capital, but we are left with the question of who, what, when, where, and how to apply this available manpower. Our challenge, designing a social system for reintegrating these "dangling particles of talent," will be to match personal interests, aptitude, and training to the mix in a way that efficiently leverages and empowers people.

During this transition period, a very real danger exists in the form of protests and repercussions from displaced workers. Those who blame their deteriorating job prospects and overall loss of opportunity on automation, could indeed wage some form of war against technology. Driverless forms of transportation will eventually supplant taxi drivers, truck drivers, bus drivers, and even airline pilots. Construction workers, craftsmen, janitors, accountants, bankers, and retailers all run a very real risk of having their positions automated out of existence.

With a combination of techno-sabotaging confrontations and pushing all the right labor-agenda political buttons, the fear of an unknown robot-infested future could take center stage as a rallying cry for new policy-setting criteria, hampering, possibly even reversing, many of the recent social advances we've made.

Even darker scenarios could play out as modern-day digital uprisings spread like wildfires, turning the speed and capabilities of tech against itself. The damage caused by a single individual could be tantamount to an anti-tech ice age spreading its influence throughout the entire world.

> "The same Internet that delivers our news and heightens our awareness of the world around us can also be used to poison people's thinking, creating an anti- technology agenda that frames the conversation for the rest of the world."

Framing the Conversation First

It's not possible for the human race to actually "run out of work." But the kind of skills needed to perform the "new work" will indeed change, and without some form of retraining intervention, the techno-illiterates run a real danger of having their prospects permanently compromised.

The re-skilling process is only as bright as the glimmers of hope and well-illuminated career path at the end of each participant's transition tunnel.

The assumption that low-skilled janitors, drivers, and dockworkers cannot be retrained for more technical work is false. It is the first of many social objections that will need to be overcome.

Rapid re-skilling programs designed to build individual competencies, one micro-capability at a time, coupled with hands-on apprenticeships and on-demand tutorial support, are all pieces of the learning environments that will be needed to elevate the caliber of workers to meet the vital workforce needs of tomorrow. Ironically, the STEM (Science, Technology, Engineering, Mathematics) talents that have prevented most of these workers from landing today's better paying jobs will be automated into the AI, artificial intelligence, operating systems of tomorrow's most ubiquitous equipment and therefore play a less significant role.

Placing Humans First

Our economy is based on people. Humans are the buying entities, the connectors, the decision-makers, and the trade partners that make our economy work. Without humans there can be no economy. So when it comes to automation:

- **A person with a toolbox is more valuable than a person without one.**
- **A person with a computer is more valuable than a person without one.**
- **A person with a robot or a machine is more valuable than a person without one.**

Automation does not happen simply for the sake of automation. Our tools do not decide for themselves what jobs and workers they will displace. Automation, like all tools, is intended to benefit people. If we only look at what automation will eliminate, we'll be viewing the world through a glass-half-empty lens. Though we have a hard time understanding the exact role of tomorrow's worker bees, even our most sophisticated machines in the future will require human owners, human controllers, human customers, and human oversight when things go wrong.

All Industries Form a Bell Curve

As with everything in life, all industry lifecycles form a bell curve with a beginning, middle, and end. It's important to understand that all industries will eventually end and get replaced by something else.

Usually the starting point can be traced to an invention or discovery, such as Alexander Graham Bell's invention of the telephone or Henry Bessemer's process for making cheap steel in large volumes. The end comes when a new industry replaces the old, like calculators replacing the slide rule. At some point along the way, every industry will experience a period of peak demand for their goods or service, after which the demand declines until the industry is replaced.

Many Industries are Entering the Downside of the Curve

Many of our largest industries today are entering the second half of the bell curve. Leading indicators that industries are entering their top-of-the-curve midlife crisis are when the disruptors, a growing cadre of startups and their process-altering technologies begin attacking key profit centers.

Prior to reaching peak demand for these goods or services, often several decades earlier, industries will experience a period of peak employment.

Peak Steel

Using "Peak Steel" as an example, the peak demand for steel is projected to occur sometime around 2024. This is when composite materials will gain enough of a foothold and the overall demand for steel will begin to decline.

Yet, peak employment for the steel industry happened in the 1970s. The 521,000 employed in 1974 were automated down to a mere 151,000 by 2000 even though the amount of steel produced tripled over that time.

In this context, any reduction in employment is a lead indicator of an industry cresting the bell curve, foretelling a downturn in the overall demand for goods or services, as an industry enters its waning years.

AS WITH EVERY 12-STEP PROGRAM, EVERYTHING BEGINS WITH ACKNOWLEDGING WE HAVE A PROBLEM. BUT THE PROBLEM TODAY IS MINUSCULE IN COMPARISON TO THE PROBLEMS THAT LAY AHEAD.

MATCHING DISPLACED WORKERS' INTERESTS WITH THE RIGHT OPPORTUNITIES FOR RETRAINING, APPRENTICESHIPS, AND JOBS WILL BE A DELICATE BALANCING ACT AT BEST.

EPIPHANY 7—
OPTIMIZING SYSTEMS

O ne of the things we use our tools for is to create *systems*.
And every one of the systems we've created, whether a planting schedule in the earliest days of agriculture, or a state-of-the-art, edge-of-tomorrow air traffic control system, is built of information.

The sheer amount of information available to us and our systems today is staggering. But it's only a small fraction of what will be available to us—and, crucially, to our *systems*—tomorrow.

All Information Ever Created Still Exists

The date is July 14, 1986. Herbert Benson wakes up in the middle of the night, goes down to the kitchen and makes himself a peanut butter sandwich. This is a seemingly trivial incident that happened nearly 30 years ago. How do we know today that this sandwich-making incident actually happened? Is there any kind of time-space record of it?

Making a nighttime sandwich is an inconsequential act, but what if Mr. Benson later committed some heinous act like killing a young girl, blowing up a bridge, or shooting the President? In these situations, whatever precedes an extremist act becomes critically important.

Was he alone in the house? Was he making sandwiches for more than one person? Did he add poison to the peanut butter? Why were there sandwich fragments at the crime scene? All of these details are critically important, but will this kind of information ever be accessible?

When we look into space we are actually looking back in time. We are looking at old light traveling towards us at 186,000 miles/second. We already know that if someone is watching us through a large telescope on the moon, they're seeing events that happened 1.3 seconds earlier—that's how long it takes light to get from the earth to the moon.

Similarly, if someone builds a giant telescope on Saturn, they will see things that happened here 75-85 minutes earlier, depending on the orbits, because that's how long it takes light to travel from the earth to Saturn. Using this as very crude proof, we already know information does indeed transcend the here and now, but can we ever access it and reassemble it into a useful form?

From what we know today, it's not reasonable to think we could send a probe 20 light-years away from the earth just to see what happened 20 years earlier. But there may be other ways to reconstruct these fragments? If people in the future somehow gain the ability to view past events, how will that change the way we live our lives? What changes would you make if you knew someone from the future might be watching you?

Knowing the Past

Every sunrise is followed by another sunset. Tides come in and wash back out. Seasons change from summer, fall, winter, spring, and back to summer again. We see trees growing in a repeatable fashion and clouds continually reforming themselves in the skies. Just as it has since the dawn of time, the metronome of life is rhythmically animating the world around us.

These are all things we associate with the movement of time, yet it is a topic we know very little about. From the standpoint of science, our understanding of time is a thimble full of wisdom in an ocean of unanswered questions.

The earth is constantly radiating information, as is every other star and planet in the universe. At the same time, every living person is also radiating information.

Throughout our lives we are spewing information of every kind in every possible direction including visual, audio, aromatic, kinesthetic, vestibular, thermoception, and countless more. With literally thousands of forms of information traveling away from our bodies every second of every day, we are left with one unanswerable question: Where does it go?

Does everything around us simply absorb these information streams? Does it bounce from one surface to another until it gets imbedded into walls, furniture, plants, and carpet? Or does it simply fade into obscurity? More importantly, is there any record of it still in existence?

My way of calculating this may seem like voodoo science, but with everything in the universe radiating thousands, possibly millions, of forms of information, the probability of at least one form still being recoverable later in time is nearly 100%. My prediction is that future scientists will someday discover the formula for unlocking the recorded history of our entire universe.

Taking this assumption a few steps further, our ability to track people and events throughout history will unleash countless business, cultural, and societal opportunities in ways that currently cannot even be imagined.

Stitching History

Going beyond traditional tools of reconstructing history—archives, diaries, and newspapers—what if we could stitch together a system that creates a viewable form of history, working with existing photo and video recordings.

When photography was first invented around 1826, no one imagined we would be snapping over a trillion photos a year, but that's exactly what has happened. At present, roughly 350 million photos are loaded onto Facebook a day.

If we assume that pictures loaded onto Facebook only represent a small fraction of the total, say 10%, that would mean we are taking 3.5 billion photos every day, or 1.3 trillion per year. Amazing as that sounds, that's probably a very low number.

The first form of photo stitching involved time-lapse photography; the act of taking snapshots a few seconds apart to capture a fluid-looking animation of specific subject matter.

Over the years, photo stitching has evolved to include the blending of side-by-side and even overlapping images into large mosaics rendered into impressively large compositions. Next generation photo stitching will involve morphing photos forwards and backwards in time using millions of data points, and millions of corollary assumptions, to fill in the information gaps. Taking photo and video stitching one step ahead, by adding other data points and information fragments, we may be able to push the time envelope even further.

Cause and Effect Relationships

In 1969, chaos theorist Edward Lorenz used the theoretical example of a butterfly's wings flapping, where that simple movement became the root cause of a hurricane forming several weeks later on the other side of the planet. This has become known as the "butterfly effect." This type of cause and effect relationship, in chaos theory, is used to describe a nonlinear system where the true sequence of events is so complex that it can only be sorted out after the fact.

But when it comes to chaos, the only chaos that really exists is our inability to comprehend the underlying cause and effect relationships happening in a hyper-complex world. Chaos is only chaos if we don't understand it. Assuming the complexities of chaos will eventually be solved, how will we leverage our newfound capabilities?

Finding Value in Old Information

Frankly, I don't know if such a system is even possible. But it's fertile conceptual territory for exploration. I've identified five initial categories for its potential use: criminal justice, spying, historical verification, biblical research, and genealogy.

The implications this technology would raise are compelling:

Would you rob a bank if you knew somebody could go back and find out exactly who did it?

Would you set someone's house on fire if someone could see you lighting the match?

Spy agencies like the NSA will naturally have a heyday deciphering the roles of participants in various forms of espionage.

If such technological advance proves to be possible, breakthroughs will come in baby steps, and pinpointing anything meaningful across time will require laser-like precision to sort through the murky universe of minutiae.

But in order to fully utilize our information resources, we must be able to access the information. Yet the adoption of each new generation, and in some cases iteration, of information technology places information stored in previous generations at risk.

This is a challenge that must be faced, starting *now*.

Protecting Endangered Information Resources

Information—in all of its forms—and formats is both our greatest treasure and most precious legacy.

How we care for, preserve, and use our vast and exponentially growing stores of information will prove one of our greatest responsibilities and challenges for the foreseeable, and especially the unforeseeable future.

Controlling Our Own Legacy

I recently attended a theatrical production of the history of my hometown of Mobridge, SD. The actors and actresses did a terrific job illustrating the tough times of the early pioneers trying to forge a new life along the Missouri River in barren lands of northern South Dakota. What I found most interesting was that this production took place in a cemetery.

The performers gave us a glimpse of the legacy left behind by these brave and bold individuals against a backdrop of tombstones and gravesites. While we know very little about those who lived 100-200 years ago, people today have the ability to leave a very detailed, well-documented legacy. In fact, they have the ability to control their reputation long after they die.

Emerging from the midst of our massive information revolution is a fascinating new industry—*legacy management*.

And one of the critical decisions each of us will have to make is whether we want to manage our legacy virtually or have it tied to a specific location.

A Growing Number of Legacy Tools

Our ability to capture snippets of our lives and preserve them has grown exponentially over the past few decades. Posting documents, photos, videos, voice recordings, and other details of our lives onto the likes of Facebook, YouTube, LinkedIn, Twitter, and Google+ has never been easier. The number of "legacy-building tools" is also growing quickly.

But at the same time, we have no good understanding of whether these tools will still exist even ten years in the future.

In short: How much of what is being captured today will still be around 500 to 1,000 years from now?

In 2005 Myspace was one of the hottest sites on the Internet, and Rupert Murdock paid $580 million to make it part of his News Corp empire. As traffic rapidly declined, Justin Timberlake and Specific Media Group purchased the hobbled company in 2011 for approximately $35 million.

In 1999 some of the top Internet properties were Lycos, Xoom, Excite, AltaVista, and GeoCities. Each of them were attracting millions of web visitors each month, competing head to head with companies like Microsoft, Yahoo, and Amazon. Today each exists in name only, resting quietly in a shadow of its former existence. It's difficult for us to think this far out when our technology is changing so quickly. Will companies like Facebook, Google, LinkedIn, and Twitter still be around one hundred years from now? Probably not.

More importantly, if companies like this disappear, what happens to all the information they collected? Organic growth often leads to organic abandonment. Is the speed with which innovative companies arrive a predictor of the speed with which they will leave?

In the midst of all these questions lie the makings of an entire new industry, one near and dear to our own hearts—building and preserving our own legacies.

As we look at the next generation of the Internet, watching carefully as it unfolds, we cannot help but be struck by how quickly connectedness has infiltrated our lives and how much of our attention it currently commands.

Much like the physical structures in our cities that form along the horizons of our urban landscapes, the data structures inside today's data giants represent some of mankind's most remarkable feats. True, they exist only as a digital compliment to the bricks and steel of physical buildings, but they hold within them vital clues about who we are, what we find valuable, and our drives and passions for forging ahead.

Building a Digital Legacy Industry

Virtually everyone wants to leave something behind. Regardless of whether it's a simple photo, a message for our great, great grandchildren, or the lessons we learned along the way, our ability to make our mark on the future is limited by our tools of preservation.

As we think through the future of information, there are three foundational pieces that will help us build this industry—finding the End of Moore's Law, the Whole Earth Genealogy Project, and Creating a Digital Preservation Culture.

1.) The End of Moore's Law—Before we can set standards for long-term data storage, we will need to find the ultimate small storage particle. Based on a piece of Moore's Law research conducted by University of Colorado's Professor Mark Dubin, we still have 129 years before we are able to store information on an individual electron. However, that ability will likely arrive much sooner as a result of the latest advancements in nanotechnology. Assuming the electron is as small as we can go, something we won't know for certain for many years to come, we can begin to set standards for future information storage. Such standards would ensure a book digitally preserved in 2150 would still be readable with technology 500 years later, in 2650.

2.) Whole Earth Genealogy Project—The genealogical industry currently exists as a million fragmented efforts happening simultaneously. While the dominant players, Ancestry.com and MyHeritage.com, have multiple websites with hundreds of millions of genealogies, there is still a much bigger opportunity waiting to happen. So far there is no comprehensive effort to build a database of humanity's heritage capable of scaling to the point of including everyone on earth, posted on an all-inclusive whole-earth family tree.

As we improve our ability to capture DNA and decipher it, it may even be possible to automate this process. The information will prove to be tremendously valuable, providing data about hereditary diseases, demographic patterns, census bureau analytics, and much more. More importantly, it will become a new organizing system for humanity—a new taxonomy. Every person on earth will have a placeholder showing exactly where they fit. In many respects, this will be similar to the way maps helped us frame our thinking about world geography. This would be a new form of "geography" for humanity, as I will show in greater detail later in this chapter.

3.) Creating a Digital Preservation Culture—While many people are rallying around efforts to "save the trees," "save our oceans," and "save our endangered species," there is virtually no effort to "save our information."

Most of the digital and analog information from only twenty years ago is unreadable with the tools and technologies we have today. Cassettes, 8-tracks, and even 3.5" disks are all becoming museum pieces as the tech world has left them as little more than a fading memory in its own digital exhaust.

One of the prized assets of today's Internet companies is their ability to amass huge volumes of digital information. But we have no provisions for preserving the data if the company itself goes under.

"Governments around the world have worked hard to create a monetary system with central banks to step in whenever a currency is failing, but we have no "central information banks" that can step in when an information company is failing."

While many still view inheritance as the primary way to leave a legacy, people now have the ability to manage, and even micro-manage, the information trail they leave behind. In fact, if they choose to, they can even communicate from here into the future with their own descendants. Archived videos, photos, and documents are only scratching the surface of what's possible here.

The body of work we leave behind has become increasingly easy to preserve. So if we chose to let future generations know who we are and why we set out to achieve the things we did, we can do that with photos, videos, and online documents.

Moving one step further into the future, generations to come will have the ability to preserve the essence of their personality and work with interactive avatars capable of speaking directly to the issues future generations will want to ask.

The digital world, even as it exists today, contains the keys to humanity, the raw essence of personhood, and in the long run, the future of our children's children. As all of us age, the notion of leaving a legacy becomes critically important, and furthering our abilities in this area will become increasingly important.

Launching the Whole Earth Genealogy Project

Of course an essential, and unavoidable, legacy is the one derived from our genetic heritage.

Some of us get bitten by the genealogical bug early in life; others get it a bit later. But there are few of us who haven't been haunted by the question—where did I come from? I've been thinking about this a lot lately. When University of Southern California researchers invented something called the Geographic Population Structure (GPS) test, which works by scanning a person's DNA for parts that were formed as a result of two ancestors from disparate populations having children, the press release instantly caught my attention.

More captivating, though, was the claim this new DNA test could locate where your relatives lived over the past thousand years, and in some cases, even pinpoint the specific village or island your ancestors came from. It's easy to draw the boxes for your own family tree going back a thousand years, but it's far more difficult finding the names, places, and detailed information about each of your ancestors.

The genealogy industry today consists of millions of fragmented efforts happening simultaneously. The duplication of effort is massive. While significant databases already exist on websites like Ancestry.com, RootsWeb, GenealogyBank, and the National Archives, there is still a much bigger opportunity waiting to happen, an opportunity to automate the creation of our genealogies.

We have the ability to create the placeholders for family trees going back a thousand, maybe even five thousand years. And now with the GPS Test we can automatically start filling in pieces of information coming from every DNA test.

Using today's stitching programs, a technology that can do pattern matching to link individual family trees whenever common names or common details show up, and using search bots to mine existing databases, we have several of the pieces already in place to begin the whole earth effort.

This kind of information becomes critically important for those looking for ways to improve personalized medicine, forensic science, and conduct research pertaining to ancestral origins of different populations. But it's far more than that. What's missing is a Jimmy Wales-type entrepreneur to turn this project into their life's calling.

The Science of Genealogy

The term Genealogy stems from a Greek phrase meaning "generational knowledge." Genealogical research is a complicated process that uses historical records and sometimes a genetic analysis to prove kinship. In

Western societies the focus on genealogy began as a way to sort out the lineage of kings and nobles who often argued about who had the most legitimate claim to wealth and power.

Some family trees have been maintained for considerable periods. As an example, the family tree of Confucius has been maintained for over 2,500 years and is listed in the Guinness Book of Records as the world's largest documented family tree. The early books of the Bible list extensive family trees for Adam & Eve, Noah, and Abraham including the lineage of Christ.

So far there has been no comprehensive over-arching effort to build a database of humanity's heritage capable of scaling to the point of including everyone on earth, going back as far as possible, to build an all-inclusive whole-earth family tree.

Missing Pieces

The Whole Earth Genealogy Project will become a new organizing system for humanity—a new taxonomy. Every person on earth will have a placeholder showing exactly where they fit. In many respects, it will be similar to the way maps helped us frame our thinking about world geography. This would be a new form of "geography" for humanity. The GPS Test will move the automation of this process a quantum leap forward, but there are still several missing pieces.

Standards—While there are a number of standards in place for conducting genealogical research and how to describe the accuracy of information and probability of kinship, there remains no consistency of layouts, file size, link strategies, or relationship formatting.

Biological Grid Map—Social networking sites create a relational grid map of personal relationships. This would be a grid map of biological relationships. The difference being that person-to-person relationships are transient, but biological ones are permanent.

The Automated Viral Piece—What's missing is a multidimensional placeholder for every person, past and present, with an automated search and contribute engine to do all the heavy lifting, that asks people to respond to their personal record.

Right to Genealogical Privacy

A recent EU court ruling has given people the "right to be forgotten", forcing Google and other search engines to remove certain links from search results. But what kind of things do people normally want to be forgotten online, and could that apply to someone's entire life, and their place in a genealogical tree?

A law giving users "the right to be forgotten" was first proposed several years ago. Google opposed the move, and through an anti-censorship campaign, has warned about the dangers of allowing people to whitewash their personal history. Employers regularly use Google and other forms of social media to check up on prospective job candidates to learn about their personal background. Negative images and posts can then be viewed, out of context, in a way that can transform something slightly objectionable into a real obstacle to getting an interview.

Could the "right to be forgotten" also be extended to the "right to mask who you're related to?" Just as many would like to amplify their genealogical links to famous people like Mozart, Edison, Mandela, Gandhi, and Einstein, others will want to disassociate their relationship to the likes of Adolph Hitler, Osama bin Laden, Ted Bundy, and Ted Kaczynski.

An Organizing System for All Forms of Life

A couple of years ago, a research team, led by Dr. Vincent Savolainen of the Imperial College of London, discovered the matK gene—what is being called the "barcode gene". They can use this gene to identify plant and animal species. As an example, they used the matK gene to identify 1,600 different species of orchid.

Combining the University of Southern California GPS Test with the matK gene for plants, we could easily begin to track genealogies for all plant and animal species as well. As I mention this possibility, I also realize the scope of a project like this suddenly scales to a level beyond comprehension. But for certain use cases, such as tracking the cause of bee colony collapse; it could reveal very useful information.

Do we have the Right to be Forgotten?

In 2001, thirty-five year old Jimmy Wales launched Wikipedia, a free, open content encyclopedia that amassed tens of thousands of volunteers to write and contribute to what has become the largest multi-language encyclopedia of all times. With Wales, it was a combination of passion, organizing skills, credibility, and a talent for managing a virtual volunteer workforce that caused Wikipedia to go viral and become successful. He was the right person, with the right talents, in the right place, at the right time.

The Whole Earth Genealogy Project will need someone with a similar passion, drive, skills, and credibility to pull off a project that will end up being exponentially larger than Wikipedia. With over 7 billion people on earth today, and over a thousand years of ancestors totaling as many as 100 billion more, the size of this project has the potential to dwarf any other existing web properties in size and scale. Every revolutionary breakthrough is just a stepping-stone to the next revolutionary breakthrough. The stars are aligning for this one now. We can't even begin to imagine the stepping-stone after this one.

"The path to a better future comes with an enhanced understanding of the past. Technology will redefine our cultures, our lifestyle, and our value systems. There will always be unintended consequences, and some of it will be difficult to manage, but new technologies will never stop coming."

Circadian Time

Of course, the system that has most clearly defined—and in too many ways dominated our lives and everything we do with them is our timekeeping system. Which is sorely in need of reinvention if not outright replacement.

Why do we hold meetings at 3:15 pm in the afternoon? The short answer is "because we can." Time, as we know it today, was invented by ancient humanity as a way of organizing the day. What began as tools for charting months, days, and years eventually turned into devices for mapping hours, minutes, and seconds.

The system I would like to see adopted is what I call *Circadian Time*.

The idea of Circadian Time started with me asking the question, "What if our clocks were oriented around sunrise instead of 12:00 noon?" What if we started every day with sunrise occurring at exactly 6:00 am? Virtually everything on earth works according to our natural circadian rhythms. Circadian rhythms involve the patterns, movements, and cycles stemming from the light and dark cycles of the earth's rotation.

Circadian Time is based on the notion that our system for timekeeping no longer has to be the rigidly pulsing overlord of humanity that constantly demands compliance. Rather, our time systems need to be oriented around people, molded to the natural flow of humanity, creating fluid structures rather than our current, abrasively rigid ones.

As I explain the following three concepts—continuous daybreak, micro-banding time zones, and virtual moments—some of the opportunities will become clearer.

Continuous Daybreak

Working with self-correcting atomic clocks, and reorienting life around daybreak, with sunrise being recalibrated on a daily basis, how would the fluctuations in the end of day change society?

In Colorado, where I live, sunrise fluctuates by nearly 3 hours between the longest days of summer and the shortest days of winter. If we were to reorient our days so sunrise would happened consistently at 6:00 am every day, then sunset would fluctuate twice as much as normal, close to 6 hours between seasons.

Admittedly, it would also be possible to orient all of our days around a 9:00 pm sunset, and push all of the daylight variations onto the beginning of each day. I'm sure some people would prefer that option. However, there is something critically important about beginning each day with the rising sun.

Perhaps a better way to orient our time structure would be to make the moment of sunrise "zero time" with our days building through a natural progression of our current twenty-four hour day. As an example, since people wake most naturally when the sun rises, work would start 1-hour later at 1:00, ending at 9:00 or 10:00.

Since I'm already introducing several new concepts, I'll refrain from dwelling on the lunacy of using 12-hour clocks to manage our twenty-four hour days. Nor will I attempt to think through the merits of metric time where we convert to ten hour days. Instead, my goal is to focus on using sunrise as the starting point for every day, and this concept is strengthened when we consider moving towards Micro-Banding Time Zones.

Micro-Banding Time Zones

Currently our time zones are structured around one hour wide geographical bands that run from the North to the South Pole. What if our current one hour time zones were reduced from broad well-defined pieces of geography with exacting borders, to one second time zones based on virtual longitudinal lines?

If we add GPS technology to our existing atomic clocks, then wherever we travel, our clocks will automatically adjust to the local time. At first blush, this sounds very confusing. With sixty seconds in a minute, and sixty minutes in an hour, and twenty-four hours in a day, that would mean we would have a total of 86,400 time zones. Yes, a crazy big number to work with, but keep in mind this is a system oriented around humans, not clocks and time zones.

Today if we schedule a call between someone in New York and San Francisco, we have to work through the mental calculation of compensating for the time zone differential between these two cities. When talking to people on the other side of the earth it becomes even more problematic with some time zones being thirty minutes or even fifteen minutes out of sync with the rest of the world. Technology could easily be developed to work with a system of 86,400 time zones, and that is where the concept of Virtual Moments comes into play.

Virtual Moments

People to people interactions are only important to the parties involved. The timing of a phone call, web conference, or virtual meeting can easily be oriented around the time that works best for the person instigating the call. Any link between two people can instantly compensate for the time differential.

Scheduling physical meetings and conferences will always take place in local time, with planning and preparation leading up to the event automatically calculated into the planning cycle. All timing will still be based on some derivative of GMT, but with far more gradients to consider and far more dependency on technological assistance.

What appears on the surface to be a massively confusing system will suddenly seem totally natural to everyone immersed in it. We suddenly become far less reliant on clocks and far more reliant on the natural order of the day.

I have no naive expectations that people will want to quickly throw in the towel on our current system for time, clocks, and time zones. We can't even decide what to do with daylight savings time, let alone invoke some massive new global change to rewrite the course of history.

That said, every avalanche begins with the movement of a single snowflake, and my hope is to move a snowflake. For people on ships, the context of time zones becomes meaningless, with an orientation of "ship time" having far more relevance to those onboard. As a way to experiment with Circadian Time, it may work well to use a floating community onboard a ship to test the theories of new social structures that will result from reinventing time.

My sense is that our current clock-centric systems are a major contributor to human health problems. We live shorter lives, produce less, and are involved in more high-stress and high-anxiety situations simply because of our rigid dedication to a time system that governs every single moment of our lives.

As a species, we can never know where our true potential lies until we confront the systems that keep us tied to the past, and that is where the true adventure will begin.

Global System Architects—Tomorrow's New Power Brokers

I often describe the future with human-like characteristics. Doing so helps me think through our relationship with the future in novel ways. Here's an example:

The future hates complacency. It hates complacency so much it has built-in self-sabotaging mechanisms to continually hold our feet to the fire. It will not allow us to shift into neutral. If we are not moving forward, we are moving backwards. There is no middle ground.

People are at their best when they are challenged. If we don't challenge ourselves, the future has a way of giving us challenges anyway. There is great value in our struggles and human nature has shown us that we only tend to value the things we struggle to achieve.

We are currently out of balance between backward looking problem-solving and forward-looking accomplishments. Forward accomplishments help erase past problems. They solve problems in a different way. We need more forward-looking accomplishments, and our greatest undertakings in the future will come in this area.

Our need for future accomplishments will also create a need for better systems to regulate, manage, and leverage the activities surrounding them. These systems will need to be global in nature, and over time, a few will emerge to challenge the power of nations. That time is coming very soon.

From National Systems to Global Systems

For decades the ringing of a bell marked the opening and closing of each day on the New York Stock Exchange. This symbolic beginning and ending of the workday set the pace for business in the US. To many, work only seemed to matter when the money people were watching. That changed once the Internet gave day-traders the ability to manage accounts around the world on the Tokyo, Honk Kong, Indian, Athens, and London stock exchanges.

Start and stop times suddenly became blurred, and eventually went away. The metronome of business began to pulse to a different beat, and the once distinct start and stop times of Wall Street ended, it created a stream of never-ending business opportunities. This ever-quickening pace of business has given strategists a whole new playbook, which has, in turn, forced companies to devise new systems and strategies to not only give them a competitive edge, but also in the process, trump the competition.

FedEx as a Global System

The epiphany moment that led to FedEx came when Yale undergraduate Fred Smith wrote a term paper based on a very simple observation: As society automated, as people began to put computers in banks to cancel checks, and people began to put sophisticated electronics in airplanes, the corporate world was going to need a completely different logistics system.

Fred was working as a charter pilot at the Tweed New Haven Airport, flying to various airports in the New England states, and talking to pilots who worked for many of the high-tech companies like IBM and Xerox, and found out what a difficult proposition it was to keep their field-service engineers and their parts and logistics systems operating. Many of the corporate airplanes had to be repurposed to fly computer and machine parts around whenever something broke down.

Several years later, after a stint in the Marines, Fred revisited the problem and discovered things had become significantly worse. Emery Air Freight was trying to solve the exact same problem using an infrastructure built around passenger planes, which weren't designed to handle freight. They were force-fitting the rapid movement of high-value-added and high technology products into a transportation system that wasn't designed for it.

So FedEx was proposed as a customized system to solve this problem. To succeed, they had to have a nationwide clearinghouse—an integrated system with trucks and planes to give the level of service that customers needed. For their network, they used the Federal Reserve Bank clearing house system, where all payments converge on a central location to complete the transaction, as their model.

That is where the Federal Express name came from. Fred wanted something that sounded substantial and nationwide, and American Express had already been taken. FedEx was made possible when the government began to deregulate the airline industry. Prior to that time, both regulation of surface and air transportation was erroneously based on linear routes, with complicated systems for making connections.

It all came together during 1977-78 when the airlines were deregulated. A couple of years later in 1980, the US federal government deregulated interstate transportation. Because the Postal Service had a monopoly on delivering mail, document delivery wasn't allowed from a legal standpoint until 1978. The standards changed in 1978 as to what constituted an item that was covered by the postal monopoly.

Merrill Lynch and a few others waged an ongoing assault through lobbyists and the press, telling Congress the USPS service was not acceptable. At the time, Merrill was being hamstrung in its efforts to move bond "Blue Books" and other types of financial prospectuses. Bowing under the pressure, congress exempted certain types of publications and documents from the postal monopoly in legislation called the Private Express Statutes.

As long as a company was delivering something overnight, and charged twice as much as the First Class postage stamps, it was exempt from the Postal monopoly. And that defined the business FedEx went into.

Backed with $91 million in venture capital funding, along with another $4 million that Fred Smith had received as an inheritance, FedEx started out with fourteen planes, initially flying between twenty-five cities. In 1985 FedEx began regularly scheduled flights to Europe, adding service to Japan in 1988, the Middle East in 1989, and the rest of the world in 1991.

In just eighteen years FedEx had become a global delivery system, unlike many of our other global systems, most of which took centuries to develop.

Eight Current Global Systems

Global systems are a fascinating area of study because they provide a context so few ever consider. When we look at early systems, such as written communications with Phoenician cuneiform, Mayan numerals, or the systems that had to be in place for engineering and building the Egyptian pyramids, it's easy to see system thinking has been around a long time, however global systems are a more recent innovation.

The most obvious advantage to global systems is the efficiencies they create. As an example, when a person who has spent their life hunting and fishing for food is able to walk to a store and purchase food, they suddenly have far more time in their life to do other things. Similarly, when a company who has had to make painful arrangements for the delivery of goods from the other side of the world can begin working with FedEx, who provides painless global delivery, the company suddenly has time to focus on other critical problems.

Here are eight examples of global systems and their development:

1. **Global Trade**—In 1264 when Marco Polo traveled the fabled "Silk Road" from Europe to what is now Beijing, China, he made some of the first inroads into creating a system for global trade.

2. **Global Sea Transportation**—On the evening of August 3, 1492, Columbus set sail under the Spanish flag from Palos, Spain, with three ships on his journey to America. This historic journey triggered an age of exploration, but more importantly gave rise to a new era of global sea transportation system.

3. **Global Measurement System**—In his 1670 book, the *Observationes diametrorum solis et lunae apparentium*, French scientist Gabriel Mouton proposed the basis for what would later become the metric system. Mouton described a decimal system of measurement based on the circumference of the Earth, creating a global measurement system recognized (although not fully adopted) by countries around the world.

4. **Global News Service**—While the telegraph was still in the early stages of development, in 1848 Paul Reuter founded the Reuters News Agency using carrier pigeons to provide the missing link between Berlin and Paris. The carrier pigeons were much faster than the post train, giving Reuter faster access to stock news from the Paris stock exchange. In 1851, the carrier pigeons were retired with the installation of a direct telegraph line. Paul Reuters played a critical role in the development of today's global news services.

5. **Global Time Zones**—In October 1884, at the request of U.S. President Chester A. Arthur, the International Meridian Conference was held in Washington, D.C. to form the basis for times and time zones around the world. Twenty-five countries were represented by forty-one delegates to establish what has become today's global time zone system.

6. **Global Air Transportation**—Charles Lindbergh, became famous for completing the first solo, non-stop flight across the Atlantic, from Roosevelt Field, Long Island, to Paris in 1927 in the "Spirit of St. Louis." This single act ushered in the age of global air transportation.

7. **Global Navigation System**—Launched in 1978, the GPS system serves as a Global Navigation Satellite System utilizing a constellation of twenty-four medium Earth orbit satellites that

transmit precise microwave signals enabling a GPS receiver to determine its location, speed and direction.

8. **The Internet**—In 1989 Tim Berners-Lee developed the Internet protocols that would become the World Wide Web, as a hypermedia initiative for global information sharing while at CERN, the European Particle Physics Laboratory in Switzerland. He wrote the first web client and server in 1990. His specifications of URIs, HTTP and HTML were refined as Web technology spread.

While these are just a sampling of the global systems that now exist, many more are on their way. In fact, the Internet has become the perfect platform for global systems to be designed, tested, and flourish.

Eight Emerging Global Systems

Here are some examples of global systems currently emerging online. I think it is safe to say none of these were started with the intention of becoming "global systems," but in the DNA of their business structures, they now exist as global systems in the making.

1. **Global Search—Google, Yahoo!, Baidu**
2. **Global Encyclopedia—Wikipedia**
3. **Global Atlas—Google Earth, Google Maps, Mapquest**
4. **Global Social Networking—Facebook, Twitter, LinkedIn, Hi5**
5. **Global Video Archive—YouTube, Vimeo, MetaCafe**
6. **Global 3-D Virtual World—Second Life, World of Warcraft, Club Penguin**
7. **Global Marketplace—eBay, Amazon, Craig's List, Buy.com**
8. **Global Music Store—iTunes, Spotify, Pandora**

Pay close attention to the nature of this list. We have just made the transition from top-down organizational structures to bottom-up organic systems that are participant driven and constantly evolving. The driving force behind developing new global systems is that each one represents a multi-*billion* dollar opportunity.

In addition to making life easier, global systems make great economic sense. The next wave of global systems, however, will not be run by corporations, but rather by a new breed of what I call Experimental Nation States, governmental-like entities that experiment with new ways for managing the world.

Eight Future Global Systems

Future global systems will emerge from today's existing industry associations. Many already have members living in multiple countries, and many seek to balance their decision-making councils with representation from each member country.

Here are eight possible future global systems:

1. **Global accounting standards for publicly traded companies**
2. **Global currency**
3. **Global airport authority to manage airport standards and policies around the world**
4. **Global oceans authority for managing everything that happens in international waters**
5. **Global genealogy systems and standards**
6. **Global ownership authority to govern standards and regulations, for personal ownership rights**
7. **Global ethics standards**
8. **Global patent systems**

Many of these organizations already exist on some level. But over time, the organizations that manage global systems will grow in influence and authority and begin to usurp the power of nations.

As an example, WIPO, the World Intellectual Property Organization is responsible for bridging the chasm between existing intellectual property laws that exist in nations around the world. Even with WIPO in place there are huge problems with competing rules, laws, and standards in the world of patents, trademarks, and copyrights.

If, sometime in the future, a person will be able to file a patent that will be universally recognized and honored around the world, the natural question becomes: What will that organization be, and what will it take for it to achieve that level of clout and authority? By its very nature, any global system aspiring to power will become a threat to existing national organizations.

The evolution of global systems will involve countless hard-fought battles against their current member base. Even though the need for global systems will be billed as a solution for the bias, fraud, and self-interest found in current national systems, the reality will be more complicated than that. In some cases, the corruption and self-interest inside global leadership teams will be greater than the corruption inside the country systems they're intended to represent. This is simply the nature of this type of authority.

Defining moments will occur when the global organizations begin to challenge the authority of their national counterparts. In some cases the organizations will formally be set up as e-democracies with members voting on every key issue.

Global System Architects—Tomorrow's Greatest Power Brokers?

One hundred years from now, what will be the most powerful entities in the world? Will it still be nations like the U.S. or will it be something else? As an example, will it be possible that corporate CEO's have far more power and control than leaders of individual countries? Will religious organizations, wielding their international clout, begin to usurp the authority of their host nations?

Will groupings of countries such as the European Union, OPEC, and the UN supersede the power of their member states? Will non-governmental organizations such as the International Monetary Fund, the World Bank, and perhaps even ICANN rise in influence to a point where they can usurp the authority of individual countries? Will the economic ties of large professional organizations, such as IEEE (Institute of Electrical and Electronics Engineers—currently with hundreds of thousands of members worldwide,) transcend the authority of the countries where their members live?

In the past, the power of a nation was considered the ability to defeat an enemy and protect its own people. But power today is more about the ability to influence and control others, even though a few still cling to the notion that power is primarily about defeating the enemy on the battlefield. In the future, a few dominant countries will continue to serve as the global police to quash uprisings and resolve disputes. But as communication systems improve, we will see fewer and fewer nations willing to openly wage war with an enemy.

Most of the power shifts between now and 2050 will result from subversive economic battles, and the ability to control or disrupt both data and revenue streams. For the disruptors, the tools for creating chaos are becoming more destructive, and soon a single individual with the right kind of gear will be able to shut down, perhaps even destroy, an entire nation.

The power centers of the future will be the countries with systems most adept at competing in the global marketplace. Large countries like China, Russia, Brazil, India, Japan, England, and USA will still play major roles, but smaller countries will have a distinct advantage derived from their ability to quickly adapt and experiment with new approaches.

So how does all this play out? My prediction is that global system architects will emerge as some of the dominant power brokers of the future, and those who recognize this shift will begin to position themselves in way to take advantage of it. A few enlightened people have already recognized this shift and the change is already beginning to happen.

IN THE FAMOUS WORDS OF JOHN F. KENNEDY, "WE CHOOSE TO GO TO THE MOON IN THIS DECADE AND DO THE OTHER THINGS, NOT BECAUSE THEY ARE EASY, BUT BECAUSE THEY ARE HARD."

CURING THE FAILING SYSTEMS THAT WILL SOON BE CRUMBLING AROUND US GIVES US A FAR DIFFERENT MOTIVATION. WE WILL BE DOING IT BECAUSE WE HAVE TO. AND YES, IT WILL BE VERY HARD.

EPIPHANY Z—
OPTIMIZING GOVERNMENT

Our governmental systems are evolving at speeds exponentially slower than the businesses, institutions or individuals that use them.

Government has virtually no competition, and consequentially, little motivation to innovate. Like a lumbering elephant alongside the sleek, cat-like speed of business, government has done little to keep pace with change, and even less to experiment, innovate, and improve.

Government workers and officials are synchronized to a radically different clock. They are neither driven to compete nor incentivized to reinvent themselves. With their change-resistant inertia firmly anchored in the past, internal government systems have grown increasingly dysfunctional, sparking a growing voice of discontent among business veterans.

Deteriorating education, increasingly expensive healthcare, a legal system with an unfathomable number of laws, seemingly corrupt financial institutions, incomprehensible tax codes, and an adversarial attitude toward change have all paved the way for a new breed of governing entities to emerge.

Entering the Era of Global Mandates

The year is 2025 and the Norwegian Nobel Committee, the organization charged with selecting the winner of the famous Nobel Peace Prize, has changed their process. They've decided to host a global election to allow the people of the world to decide which of the candidates is the most deserving.

Two months before the election, a slate of four possible candidates is announced. The election itself takes place over a twenty-four hour period and a secure online voting system allows voters to make their selection from any computer, tablet, or cellphone.

As a way to push voters to learn more about each contender, they are given a short test consisting of eight simple questions, two about each candidate, before the official vote is cast. Peace advocates around the globe are anxious to participate, and once Election Day formally clicks to an end, a total of seven hundred and forty million voters from fifty different counties have selected the winner.

With spotlights blazing and countless news cameras poised to capture the moment, the winner is formally announced. However, unlike previous winners, this person suddenly becomes the most famous person in the world, more celebrated than any king, president, or prime minister on the planet. While on the surface this may appear to be nothing more than an ingenious PR stunt for selecting prizewinners, it is indeed much more. Voting software that crosses country lines falls into the category of "catalytic innovations" with the potential of creating new global mandates.

Catalytic Innovation

Unlike disruptive innovation that disrupts an existing industry, catalytic innovation has the potential to spawn entirely new industries. Any technology that becomes a catalyzing agent for opening doorways into a world never before seen, falls into the broad new category of catalytic innovation. Electronic voting has long been touted as a more efficient way to conduct an election. But the evolving technology will soon offer far more than just a new face to an old process.

Possible Scenarios

Early forms of global elections are already in play with shows like *American Idol* and the *Eurovision Song Competition*. Moving past the music scene, in what situations do global elections become an appropriate tool for influencing the future?

Certainly there are many relatively harmless scenarios that could be envisioned:

- **Selecting the *Time Magazine* "Person of the Year"**
- **Determining the location of the next Olympics or World Cup**
- **Re-designating the official "Seven Wonders of the World"**

There are also many ways a global election could be overreaching, pushing past our present limits of acceptability. Here are a few examples global elections that would likely get censored out of or disallowed in many countries:

- **Selecting the official leader of the world**
- **Deciding who officially owns the moon**
- **Attempts to globally override the results of an individual country's election**

- **Designating an official religion for the world**

At the same time there are huge grey areas where the appropriateness of this kind of election is unknown.

- **Should there be a global ban on plastic bottles that are not biodegradable?**
- **Should we place limits on the amount of fishing, mining, pollution, or deforestation that takes place around the world?**
- **Could there be a global code of ethics, bill of personal freedoms, laws of right and wrong, privacy policies, etc.?**

Many of our national systems are becoming part of new global systems. As such, we could use global elections to determine the official policy for:

- **The Internet**
- **International airlines**
- **GPS—Global positioning systems**
- **Global currencies**
- **Intellectual property**
- **Global passports**
- **Central banks**

Many people have predicted we will someday enter an era of e-democracy where citizens can weigh in on far more issues than those that end up on a ballot.

Since it circumvents the power of elected officials, e-democracy will face considerable resistance early on. But resistance will increase considerably as it extends beyond country borders disrupting the influence of more than a few local power brokers.

Global Mandates

In the context of global elections, at what point will the results be large enough and representative enough to influence, even supersede, the authority of an industry group, individual country, or even a group of countries?

Would the participation of twenty million people from five countries constitute a global mandate? Probably not.

How about 500 million people from eighty countries? Or would it require a certain percentage of the world's population, say ten percent, which translates into 700 million people?

Every industry or topical area has its own "affected population." Should the vote require a minimum percentage of an affected population? If the results of an election were only separated by one percent of the vote, is that still too close to be considered a global mandate, or will it require some sort of "obvious will of the people" supermajority such as 60-70%?

At this point, there are far more questions than answers. But that will not stop people from using global elections to influence the world.

The Coming Battle of the Mega-Influencers

There are several people who have the ability to sway world opinion. As an example, if a global election were being orchestrated by someone like Mark Zuckerberg, Jack Ma, Larry Page, Jimmy Wales, Mark Cuban, Reid Hoffman, Marissa Mayer, or Peter Thiel, most of the whole world would take notice.

Companies like Facebook or Google, with over a billion users worldwide, may be the perfect platform for both promoting and conducting early elections and testing the limits, but these are not the only ones.

While the motivations behind a Facebook or Google-led election may seem completely altruistic on the surface, their behind-the-scenes plan may be to influence heavily censored countries like China, coercing it to open its doors to tech companies or face being "left out" of key global decisions. Any election suspected of having a hidden agenda will be destined to fail.

Leveraging the Internet for Influential Purpose

So far we've only scratched the surface of how the Internet can be leveraged to influence the entire world. Pushing beyond the current playgrounds of photo and video sharing, online games, and social networks are apps with far reaching implications. Electronic voting is one of them.

People of power and influence who want to test the limits will orchestrate the first wave of global elections. Most of the first wave of elections will be poorly conceived, attract a "too-small-to-notice" following, and will be considered by most to be a failure. But much like Peter Diamandis' effort to catalyze global prize competitions, someone will emerge as a thought leader to pioneer a new global election industry.

Gestating inside what may start out as a playful way to get more people involved in selecting the winner of the Nobel Peace Prize is actually something far more serious—methodologies for establishing new global mandates. In best-case scenarios, global mandates will serve as everything from a temperature gauge for global opinion to a checks and balance system for over-reaching countries.

When Prisons Become Illegal

The first time I watched *Star Trek* and heard Captain Kirk utter the phrase—"Set your phasers to stun!"—it occurred to me these future weapons featured a number of different settings. While most people assumed a simple two-position switch with only "kill" or "stun" options, I found myself dwelling on the possibilities of an eight or ten-position switch and wondering what the other options might be.

Perhaps they would include stun 1 (with pain), stun 2 (without pain), giggle (make them laugh uncontrollably), amnesia (forget what they're doing), slo-mo (causing them to move in slow motion), suicide (making them take their own life), seizure (all muscles fire at once), overwhelming guilt (immobilized by guilt and self-loathing), or overwhelming pity (suddenly they become your friend).

Since today's weapons have one setting, we have a hard time imagining a weapon that offers a variety of choices.

Similarly, when people show up in court, judges only tend to have one setting for punishment—incarceration. While judges do have more options, like setting fines and community service, with our existing infrastructure built up around jails and prisons, we have a justice system that has a hard time considering other options.

With incarceration rates in the U.S. now reaching epidemic levels, I would like to take you through the exercise of envisioning a world where prisons are no longer an option. If judges no longer had 'incarceration' as a setting on their gavel of justice, what kind of world would we live in? Here are a few thoughts:

Our Current Incarceration Epidemic

The United States represents only five percent of the world population yet houses twenty-five percent of the global prison population. Nearly half of all prisoners in the USA are incarcerated for non-violent crimes. During 2011, the total U.S. prison population declined for the second consecutive year, to under 1.6 million inmates or 15,023 fewer inmates than in 2010. This represents a 0.9 percent decrease in the total prison population.

The positive trend, however, is still only a small blip in a much larger problem. Statistics show the U.S. prison population rose by 708% from 1972 to 2008, a rate far outpacing that of general population growth and crime rates. While one out of every 122 Americans is now actually in prison, one out of every thirty-two of us is either in prison or on parole.

This means that seven million adult men and women—about 4.1% of the total U.S. adult population—are now very involuntary members of America's "correctional community." According to the U.S. Bureau of Justice Statistics the United States still has the highest documented incarceration rates and most overcrowded prisons in the world.

Punishment Vs. Rehabilitation

Our current criminal justice system has been developed around several distinct stages, including arrest, prosecution, trial, sentencing, and punishment, which usually involves some form of imprisonment. This is a very expensive process that employs millions of people throughout the entire system. Rehabilitation is the idea of transforming a person who has committed a criminal act so they won't do it again.

Rehabilitation can take place at any stage of the system. When a police officer makes first contact with a criminal, they can impose on-the-spot penalties and warnings.

It is in the sentencing and punishment stages of the process most controversy lies. Advocates of the status quo argue there is great value in punishment as a deterrent and even greater value in removing disruptive personalities from society.

The counter-argument is that punishment actually leads to greater degrees of criminality, and prisons themselves become something akin to a college for crime, with many people leaving prison worse than when they went in. While good arguments can be made on both sides, there is little argument over the cost of maintaining the current system.

Not only are taxpayers picking up the cost of police, attorneys, judges, and prisons, but hidden many layers below is the overall toll this system exacts on society. Prisons separate wage earners from their incomes, parents from their children, and self-sufficiency from state support. In a paycheck-to-paycheck society, incarceration almost always leads to bankruptcy, and a "felony" offense on your criminal record most often leads to a life of poverty.

A Note About Restorative Justice

Restorative justice is an approach to justice that focuses on the needs of the victims and the offenders, as well as the surrounding community. Crime is a larger issue than just how government deals with criminals. Since crime victims and the community bear the brunt of the crime, they too must be actively involved in the criminal justice process for a true change to occur.

Currently, many victims feel re-victimized by the criminal justice system itself because it excludes them from most of the process. Using a different approach, restorative justice victims take an active role in the process, while offenders are encouraged to take responsibility for their actions, "to repair the harm they've done—by apologizing, returning stolen money, or community service".

Restorative justice asks that victim's concerns be considered throughout the process.

Victims need help regaining a sense of control over their lives, and they need to be compensated for their loss. Rather than simply warehousing offenders, restorative justice system holds offenders personally accountable. They need to confront the pain they have caused to their victims and take the steps necessary to overcome their criminal behavior.

While restorative justice offers a positive step in the right direction, let's take some of these ideas a few steps further.

Envisioning a World Without Prisons

Let's start with the simple question, "If prisons were removed as an option for working with criminals, how would we deal with them?"

In much the same way we considered different settings on a *Star Trek* phaser, how could we develop different ways of managing criminal offenders? As technology progresses, here are a few possibilities:

- **Automated Monitoring:** With drone technology progressing quickly, we will soon have the ability to assign a small, inconspicuous, hovering drone to 24/7 surveillance on an individual. Over the next few years drones will be able to lock onto a specific digital signature (perhaps a heat signature combined with tracking brain and energy waves) for every individual. Once we can track an individual, the next step will be to auto-analyze the person's actions and take measures instantly whenever a violation occurs.

- **Automated Correction:** As we move into an increasingly transparent society, deviant behaviors will be increasingly easy to detect. As an example, pocketing merchandise in a store, breaking into a house, or hitting/killing someone will be very easy to track. For someone who is a habitual offender, the same drones that are used to monitor someone can be used to dispense warning shocks, Taser jolts, or other forms of stopping the aggression.

- **Automated Punishment:** When dealing with more violent individuals, people could be sensed to "electronic canings" where a drone randomly shocks someone a designated number of times over a certain period of time. One caveat for resorting measures like this would be that the punishment couldn't put the person in harm's way or cause problems for their family or friends.

- **Punishment Matching the Crime:** In the future we will have the ability to painlessly alter the function of the human body to closely match the punishment to the crime. As an example, a person who is a pickpocket will lose their ability to use their right arm for the next six years, or a burglar will lose their ability to use their legs for the next eight years, or a rapist will experience a constant state of erectile dysfunction for the next twenty years.

- **Brain Wipes:** In extreme cases, the only viable option may be to wipe the brain clean and start over, but when we force people to relearn how to walk, talk, and feed themselves all over again, we can only begin to think through the moral and ethical implications of a brain wipe.

The advantage of most of these approaches is the offenders can still remain productive members of society and the costs will be trivial compared to incarceration today. Yes, technologies like these will lead to many abuses. But as we peel away the onionskins of transparency and add new layers of technology, our options for auto-monitoring, auto-correction, and automated forms of punishment will yield unusual new alternatives for righting the wrongs of deviant people in society.

All Those Damn Laws! Over 18 Million Laws in the U.S. and Climbing

How many laws are affecting you as you're reading this today?

If you think you know the answer, I will tell you you're lying, and there is a law against lying about how many laws there are.

I really don't know that there is such a law, but then again, *you* really don't know that there isn't. So we could *both* be in trouble.

I do know ignorance of the law is no defense. This is something I've heard many times in the past, but I have no clue as to whether it's really law or just something judges use to belittle people into feeling guilty.

Thinking through the title for this section, I have to admit I have no idea how many laws exist in the U.S. But then again, neither does anyone else.

They're simply not countable. There is no central repository for our laws, no common form, style, or accessibility requirements; only some level of hope that once laws are enacted, people will pay attention to them.

Here's why this is such a confusing issue.

The total number of governmental bodies in the U.S. is approaching a staggering number—90,000. Every city, county, state, and special taxing district has its own governing body with its own elected officials.

Taking on many of the characteristics of living, breathing organisms, these governmental organizations are constantly fighting for influence, control, and survival.

Each one of these governmental entities has an ability to create and enforce its own laws, rules, and regulations. Working with a limited set of tools in their toolbox, governments have resorted to using laws and regulations to solve virtually every conceivable problem. The sheer volume of laws emerging from these 90,000 rule-making bodies is truly stunning.

It may indeed total eighteen million, a number that was mentioned to me by a friend, but has no real basis that I've been able to uncover.

With a society that is already heavily invested in our current systems, and people already pre-programmed to think and act accordingly, what we need is a system for changing the system.

Here's what I propose.

Society's Operating System

Much like a computer operating system, our body of laws serves as the code for all citizens to abide by.

From a computer nerd perspective, writing a computer program that uses eighteen million lines of code to accomplish the same thing as one with 1,800 lines would be considered a massively bloated program.

That's exactly what's happening with our laws. As an operating system, they demand far too much human energy and intellectual bandwidth to keep each of these fiefdoms running.

Making matters worse is the lack of any central repository for our laws. Some only exist on scraps of paper stored in filing cabinets in courthouses, while others have been meticulously stored in books and other digital medium.

Using another computer analogy, the lack of a central repository is like trying to operate a computer without a central cache for its memory.

This leaves us with a very dysfunctional operating system, and the only way to change an operating system is to rewrite the source code.

Proposing a Solution: Laws That Manage the Laws

We currently have no check-and-balance for impeding the excessive law writing now taking place.

For this reason, I would propose a four-step system for correcting the system. These are what I refer to as the "four laws for managing the laws:"

1. **Public Access Requirement:** Make it a requirement that all laws be posted in one central online location—one central website for all laws. Any laws not posted on this website will be deemed unenforceable.

2. **Sunset Provision:** Any laws that have not been applied or enforced in that past twenty years become unenforceable and

must be removed from the list. Time spent getting rid of the clutter means less time for creating new laws.

3. **Simplification Mandate:** All laws must be written on an eighth grade comprehension level. No laws can go into effect until they are certified as having been written on this level.

4. **Code of Government Ethics:** No governmental entity will be allowed to directly profit from the enforcement of its own laws. The control of wealth is just as insidious as taking ownership of it. Whenever there's a direct profit motive linked to law enforcement, the nature of government changes, and our humanity becomes compromised.

If I were to add a fifth requirement, it would be that all new laws be game tested prior to implementation.

For many, the process of modeling and game testing our systems is a cause with epic meaning, something many would want to participate in.

Game designers would love the challenge. Game players will enjoy being part of something far bigger than themselves. Even politicians would love it because it gives them a logical path for answers.

However, taking these steps is only part of the answer.

Reigning in the Unreignable

Technology ends up being the great enabler of complexity. Is it possible for technology to take our existing super complex set of laws and turn them into something manageable, even reasonable?

If all of the laws reside in a central place, we can develop artificially intelligent systems that can read, understand, and know how to apply them. Smartphones and other AI devices can let us know when we're in a gray area or about to violate a law.

Applying machine learning to our courts and justice systems will finally make the phrase, "ignorance of the law is no defense," a viable concept.

We won't need to personally know the laws; our devices will do that for us. They will serve as our guide, our coach, and in some respects, even our conscience.

Is it reasonable to assume morality can be automated, that our societal norms and human faux pas can be reined in?

Does that mean, by extension, automating a system for managing and enforcing all our laws will make us dysfunctional? Or will it simply make us more efficient?

Does the notion of having a machine that can tell us the difference between right and wrong scare you?

The rogue philosopher in me says this is the worst idea ever.

But at the same time, the entrepreneur in me thinks there may be a golden opportunity for turning the world's worst idea into something exceptionally great.

EVERY COUNTRY ON THE PLANET IS ABOUT TO UNDERGO HEIGHTENED LEVELS OF SCRUTINY, BOTH INTERNALLY AND EXTERNALLY AS OUR DIGITAL AWARENESS GROWS. THIS WILL, IN TURN, FORCE GOVERNMENTS TO RETHINK VIRTUALLY EVERY SYSTEM, PROCESS, AND STRATEGY AS IT RELATE TO THEIR CITIZENS.

EPIPHANY 7— OPTIMIZING INFRASTRUCTURE

E ven as our 21st Century world evolves at an accelerating pace, much of our lives are spent dealing with the drawbacks and increasing collapses of 20th Century—and more 19th Century than we'd care to imagine—infrastructure.

Yet the calls to rebuild that infrastructure, to spend billions or even trillions on "shovel-ready" projects are becoming even more challenging to fund.

The Curse of Infrastructure

Every time I drive to the office there are eleven separate stoplights along my route. Based on some cosmic luck-of-the-draw, two thirds of the stoplights will either be red or green, and the time it takes me will vary from twelve to twenty-two minutes. It's a relatively short commute.

But the countless hours spent every year sitting mindlessly at ill-timed stoplights represents a tremendous expense of time, fuel, and resources that not only I, but also the majority of workers in America, bear, all because of one tiny piece of ancient infrastructure—the dumb stoplight.

Indeed, many communities are beginning to shift to intelligent traffic systems that constantly adjust patterns to better match the flow of cars. But this long overdue transition is happening at great expense to cities, an expense that cities themselves derive very little direct benefit from. In this one teeny example, we can begin to see the challenges ahead for dealing with infrastructure.

Stoplight systems are not only expensive to maintain and upgrade; more importantly, the existing system blinds us to *what-comes-next*. For this reason, I'd like to take you along on a journey into the complex world of future infrastructure, and the curse of every legacy system that accompanies it.

Working Within the Current System

Clearly I'm not the only one complaining about stoplights. A group of researchers from MIT and Princeton University have developed an app that takes advantage of a growing trend: drivers who install brackets on their dashboards and mount their smartphones as GPS navigators.

The researchers used a network of these GPS-enabled cell phones to collect information about traffic lights. Based on images captured by the phones' cameras, the app is able to predict exactly how slowly a person needs to drive in order to miss the next red light.

The app, called SignalGuru, was tested on twenty cars in Cambridge, Massachusetts and Singapore. The system used in Cambridge, where lights change according to fixed schedules, predicted the change of red lights to within two-thirds of a second. In Singapore, where traffic lights change depending on traffic flow, the system was less precise.

This is an example of people going to extreme lengths to compensate for our ailing infrastructure.

Philosophy of Infrastructure

There is a long-held belief that infrastructure in general represents a long-term societal investment that will move us along the path of building a more efficient, better functioning, society. And usually it does… for a while.

However, infrastructure comes in many forms, and as we build our elaborate networks of pipes, wires, roads, bridges, tunnels, buildings, and waterways, we become focused on the here and now, with little thought as to whether there might be a better way.

1. **Once wired power lines are put into place, it becomes hard to imagine us using wireless power.**
2. **Once a human-based delivery system is put into place, like the post office, it becomes hard to imagine a human-less automated delivery system.**
3. **Once a tunnel is bored through a mountain, it becomes hard to imagine a better way to get to the other side.**
4. **Once a prison is built, it becomes hard for us to imagine a prison-less justice system.**
5. **Once an airport is constructed, it becomes hard to imagine air transportation in any other way.**
6. **Once a highway is built, it becomes hard to imagine an alternative transportation system that uses something else.**

Infrastructure creates its own inertia. As soon as it's in place we suddenly stop thinking about what comes next. Our life is based on stories of the here-and-now.

Once stories are told, it becomes hard to un-tell them.

Sacred Cow Syndrome

In many respects, infrastructure becomes a lasting testament to who we are as a society and part of the cultural moorings we use to guide our existence. People become emotionally invested in them because they create stability, usefulness, and purpose. But more importantly, people become financially invested in them and their livelihood depends on their ongoing existence.

Virtually every piece of infrastructure creates jobs, revenue streams, and investment opportunities, as well as new laws, regulations, and industry standards.

The longer a piece of infrastructure is in place, the greater the resistance to replacing it. Much like an aging tree, the root system that feeds it becomes enormous. The world of infrastructure has far too many sacred cows with built-in inertias that are highly resistant to change.

Life Cycles Are Getting Shorter

Whenever a new piece of infrastructure is put into place, the clock starts ticking. The corrosiveness of nature, structural deterioration, and functional obsolescence all begin to rear their ugly heads. An infrastructure's useful life may be measured in decades or in centuries, but all forms of infrastructure will eventually wear out.

Life cycles for virtually all forms of infrastructure are getting shorter. On the long end of the spectrum, many of the hydroelectric dams in the U.S. were built in the 50s and 60s. But with modifications and upkeep, these dams still have many useful decades ahead of them.

Lasting considerably less time, the usable life of shopping centers is around ten years before major renovation, and often less than twenty years before they're torn down completely. Similarly, experts are now viewing the usable life of large stadiums shortening from fifty years to somewhere around twenty years.

Eight Stages of the Curse

As with most of the cycles we deal with in life, there are well-defined stages that infrastructure goes through during its existence.

1. **Celebration**—Once a new project is complete, we begin by patting ourselves on the back in celebration of this latest accomplishment.
2. **Acceptance**—It usually doesn't take long for people to accept it and make it part of their daily life.
3. **Dependence**—Over time we lose sight of what life was like without it and we learn to rely on it as a routine part of life.
4. **Deterioration**—All man-made structures eventually wear out, and once they do, we look for something new and better to replace them.
5. **Disagreement**—Repair is almost always cheaper than replacement, and the vocal few that have their eyes on something better, have to wait.
6. **Denial**—With ongoing repairs being made, it becomes easy to deny any problem exists.
7. **Agonizing Sunset**—Even when up to date, better systems are being used elsewhere, the replacement decision will drag on, and on, and on.
8. **Painful Transition**—Eventually the replacement decision will come, but it will come at a price. Change is never easy to accept, especially when countless numbers of individuals become heavily invested in the surrounding systems.

As you are starting to see, our aging and problem-riddled infrastructure, and the painfully slow processes we have for changing it, is taking its toll. Here's one example:

Disruptive Infrastructure

As with other areas of society, disruptive technologies are beginning to attack the sacred cows of infrastructure. Here are a few fascinating projects, some of which we've looked at in other contexts, where the founders are working hard to disrupt the status quo.

ET3: Billed as "Space travel on earth," ET3 is an Evacuated Tube Transport Technology that transports packages or people inside car-sized capsules on a frictionless maglev track that moves effortlessly inside airless tubes to the destination. It's estimated that a speed of up to 4,000 mph can be achieved with this system.

Eole Water: The Eole Water Company developed a special wind turbine capable of extracting upwards of 1,000 liters of water a day directly from the air.

Contour Crafting: Developed by USC Professor Dr. Behrokh Khoshnevis, contour crafting is a layered fabrication technology similar to 3D printing that can be used to construct buildings and other key pieces of infrastructure.

Google's Driverless Cars: After conducting over one hundred separate test and logging over 250,000 hours on their fleet of driverless cars, Google is setting their sights on a very disruptive system that will forever change transportation. Hidden behind the hype of this technology is Google's plan to come up with an Android-like operating system for all future driverless cars. As cars become driverless, we will see dramatic shifts in how roads and highways are built.

Blueseed: Funded by PayPal founder Peter Thiel, Blueseed proposes to create visa-free floating work villages in international waters, with the first to be located within helicopter distance of Silicon Valley. This floating island is a new form of infrastructure destined to disrupt a variety of existing systems.

Airdrop Irrigation: Winning the James Dyson Award for Innovation, Edward Linnacre has developed an inexpensive self-contained solar-power irrigation unit capable of extracting water from the air to add moisture to surrounding plants.

Bitcoin: Even though it's a virtual currency with fluctuating value, many investors now think Bitcoin is safer than the euro, and it can be used for real-world purchases. The advantage of Bitcoin as a currency is that it is decentralized, and an algorithm, rather than a bank or a government entity controls the supply of Bitcoins.

Much of the world around us has been formed around key pieces of infrastructure. In spite of its tremendous value, infrastructure is expensive to maintain, hard to change, and generally limits how we think about the future.

Eventually change will happen, but people who are at the heart of these changes pay a price. Transitions like this can be very painful. That said, the lifecycles for infrastructure are getting shorter, and the teams driving the disruptive technologies are getting far more sophisticated.

"Infrastructure projects represent huge paydays for someone; usually many someones, and the disruptors are determined to make it their payday. By 2030, we will see more changes to core infrastructure than in the combined total in all of human history."

Our sacred cows are about to be set free, and the fundamental shifts we will see to the way society functions will be nothing short of breathtaking.

2050 and the Future of Infrastructure

The life cycle of infrastructure is getting shorter, and teams driving the disruptive technologies are getting far more sophisticated.

Here are ten examples of how our core infrastructure is about to change and what this will mean to the nations and businesses at the heart of this revolution.

1.) Driverless Cars and Driverless Highways

Even though the art of road building has been continually improving since the Roman Empire first decided to make roads a permanent part of their infrastructure, highways today remain as little more than dumb surfaces with virtually no data flowing between the vehicles and the road itself. That is about to change.

Driverless technology will initially require a driver, and it will creep into everyday use much as airbags did, first as an expensive option for luxury cars, but eventually as a safety feature required by governments. The greatest benefits of this kind of automation won't be realized until the driver's hands are off the wheel.

With millions of people involved in car accidents every year, it won't take long for policy-makers to be convinced driverless cars are a safer option.

The privilege of driving is about to be redefined. As cars become equipped with driverless technology, important things begin to happen. To compensate for the loss of a driver, vehicles will need to become more aware of their surroundings.

Using cameras and other sensors, an onboard computer will log information over a thousand times per second from short-range transmitters on surrounding road conditions, including where other cars are and what they are doing. This constant flow of data will give the vehicle a rudimentary sense of awareness.

With this continuous flow of sensory information, vehicles will begin to form a symbiotic relationship with their environment, a relationship far different than the current human to road relationship, which is largely emotion-based. An intelligent car coupled with an intelligent road is a powerful force. Together they will accelerate our mobility as a society, and do it in a stellar fashion.

- **Lane Compression**—Highway lanes need only be as wide as the vehicles themselves. Narrow vehicles can be in very narrow lanes, and with varying sizes and shapes of vehicles, an intelligent road system will have the ability to shift lane widths on the fly.
- **Distance Compression**—With machine-controlled vehicles, the distance between bumpers can be compressed from multiple car lengths to mere inches.
- **Time Compression**—Smart roads are fast roads. Travel speed will be increased at the same time safety is improved.

In the driverless era, intelligent highways will be able to accommodate fifty to one hundred times as many vehicles as they do today. Counter to traditional thinking about vehicle safety, the higher the speeds, the fewer the number of vehicles on the roads at any given moment.

As we compress the time and space requirements of every vehicle, we achieve a far higher yield of passenger benefits per square meter of road resources.

In addition to the benefits passengers receive, the road itself will greatly benefit from this technology. With cars constantly monitoring road conditions, the road itself can call for its own repair.

Rather than waiting until a pothole or other problem becomes a serious hazard, as is currently the case, and repair crews disrupt traffic for hours, days, or longer, micro repairs can happen on a daily, sometimes hourly, basis. High-speed coatings and surface repairs can even be developed for in-traffic application. Even treacherous snow and ice conditions will have little effect if de-icer is applied immediately and traffic is relentless enough.

2.) Tube Transportation Networks

When Tesla Motors CEO, Elon Musk, mysteriously leaked that he was working on his Hyperloop Project, the combination of secrecy, cryptic details, and his own flair for the dramatic all contributed to the media frenzy that followed. Leading up to this announcement was his growing anxiety over California's effort to build a very expensive high-speed rail line between Los Angeles and San Francisco with outdated technology.

While the Musk media train was picking up steam, several reporters pointed out a similar effort by Daryl Oster and his Longmont, Colorado-based ET3, to build a comparable tube transportation system that was much further along.

3.) Atmospheric Water Harvesters

Only two percent of the world's water is fresh. To make matters worse, only one-fourth of all fresh water is accessible to humans.

Until now, the entire human race has survived on 0.5% of the available water on earth. But that's about to change.

We are seeing a fast growing trend towards harvesting water from the atmosphere, something our ancestors first began working on centuries ago. People in the Middle East and Europe devised the original air-well systems over two-thousand years ago. Later the Incas were able to sustain their culture above rain line by collecting dew and channeling it into cisterns for later use.

Even though these techniques have been around for a long time, technology in this area has recently taken a quantum leap forward, and many are beginning to think in terms of houses that generate their own water supply, self-irrigating crops, and even "waterless" cities. The

Earth's atmosphere is a far more elegant water distribution system than rivers, reservoirs, and underground waterways.

Our current systems involve pipes and pumping stations that are expensive to operate and maintain, and easily contaminated. There are roughly 37,500 trillion gallons of "fresh" water in the air at any given moment. The age-old problem has been getting it to people who need it at exactly the right time.

A new breed of inventors has emerged to tackle this exact problem. Using solar, wind, and other forms of passive energy, our future water networks will operate with far more efficiency and convenience than anything imaginable today. Today's steel pipes will soon be replaced with tomorrow's air pipes, and we will forget what life was like when chlorine-tasting water was an everyday occurrence.

4.) Micro Colleges

When Facebook announced the $2 billion acquisition of Oculus Rift, they not only put a giant stamp of approval on the technology, but they also triggered an instant demand for virtual reality designers, developers, and engineers. Virtual reality professionals were nowhere to be found on the list of hot skills needed for 2014, but they certainly will be for 2020.

The same was true when Google and Facebook both announced the acquisition of solar powered drone companies Titan and Ascenta respectively. Suddenly we began seeing a dramatic uptick in the need for solar-drone engineers, drone-pilots, air rights lobbyists, global network planners, analysts, engineers, and logisticians.

Bold companies making moves like this are instantly triggering the need for talented people with skills aligned to grow with these cutting edge industries.

Whether it's Tesla Motors announcing the creation of a fully automated battery factory, Intel buying the wearable tech company Basic

Science, Apple buying Dr. Dre's Beats Electronics, or Google's purchase of Dropcam, Nest, and Skybox, the business world is forecasting the need for radically different skills than colleges and universities are preparing students for.

In these types of industries, it's no longer possible to project the talent needs of business and industry five to six years in advance, the time it takes most universities to develop a new degree program and graduate their first class. Instead, these new skill-shifts come wrapped in a very short lead-time, often as little as three to four months.

With literally millions of people needing to shift careers every year, and the long drawn out cycles of traditional colleges being a poor solution for time-crunched rank-and-file displaced workers, we will be seeing a massive new opportunity arising for short-term, pre-apprenticeship training in the form of Micro Colleges.

5.) Space-Based Power Stations

Earth's appetite for power continues to grow. Since the 1960s, power consumption has quadrupled around the globe, with many countries opting to build large oil and coal plants to meet the demand. For Japan, a burgeoning economy without large oil and coal reserves, after the Fukushima disaster occurred, an in-depth review concluded the most viable long-term strategy was to focus on spaced-based power systems.

For this reason, the Japan Aerospace Exploration Agency (JAXA) recently announced its twenty-five year plan to build the world's first 1-gigawatt power plant in space. The vision of harvesting solar power from space and beaming it to earth has been around ever since Dr. Peter Glaser first proposed it in 1968.

After considerable research in the 1970s, scientists concluded it wasn't a viable concept just yet because technology hadn't advanced enough. The materials were far too heavy, and it would have required over a hundred astronauts working with thousands of crude robots to create it.

Since then, technology has advanced in countless ways, not only making it doable, but for Japan, making it the best available option for controlling its own destiny. What most people don't realize is that solar panels in space are ten times more efficient than those on earth because there are no day-night cycles, seasonal variations, or weather issues to contend with.

Here's where it gets even more interesting. Many other countries won't be comfortable with Japan having the world's only expertise in building space-based power stations. Once the first one proves successful, it will become faster and cheaper to launch the next ten, or even one hundred of them.

6.) Drone Delivery Networks

According to the Association for Unmanned Vehicles International, once drones get approved for commercial use, the first three years will produce a multi-billion-dollar industry employing hundreds of thousands of new manufacturing jobs. More than just manufacturing, there will be a need for drone pilots, drone farming specialists, drone security, drone data analysts, drone mosquito killers, and much more.

7.) Mass Energy Storage

We are now entering the early growth stages of what will surely become a huge global industry—energy storage. It will both support and compete with conventional generation, transmission and distribution systems.

Over the coming decade as the industry evolves, it will lead to new business models and the creation of new companies that make, apply and operate storage assets to help the grid work more reliably and cost-effectively, while decreasing unwanted environmental impacts.

8.) Global Language Archive

For most of us, the language we speak is like the air we breathe. But what happens when we wake up and find our air is going extinct? Researchers estimate that over the last five hundred years, half of the world's languages, from Etruscan to Tasmanian, have vanished.

By the next century nearly half of the roughly seven thousand languages spoken on Earth will disappear, as young people abandon native tongues in favor of English, Mandarin, or Spanish. Think of the Global Language Archive as the "Louvre of Languages" where culture and language collide in a way that all can experience.

9.) Whole Earth Genealogy Project

The genealogy industry today consists of millions of fragmented efforts happening simultaneously. The duplication of effort is massive. While significant databases already exist on websites like Ancestry.com, RootsWeb, GenealogyBank, and the National Archives, there is still a much bigger opportunity waiting to happen, an opportunity to automate the creation of our genealogies.

10.) Our Trillion-Sensor Infrastructure

In the last six years, we've gone from ten million sensors—in things like the Nintendo Wii and iPhones—to 3.5 billion. This is why Janusz Bryzek, an executive at Fairchild, organized the Trillion Sensor Summit. Bryzek is projecting 1 trillion sensors by 2024 and 100 trillion sensors in the mid 2036 along with literally millions of new primary and secondary jobs to manage this emerging sector.

The projects I've listed here merely scratch the surface of what's possible. Whether it's building the Great Pyramids in Egypt, erecting the Great Wall of China, or sending someone to the moon, crazy-big projects have a way of defining our humanity and raising the bar for future generations.

As our capabilities improve, we simply need to set our sights higher and aim for the stars… literally!

By 2050, we will see more changes to core infrastructure than in the combined total in all of human history. The fundamental shifts we will see to the way society functions will be nothing short of breathtaking.

For the balance of the chapter I'll be examining in some detail one of the coming infrastructure projects. But as you read about tube transport, bear in mind it *is* only one of many world-changing projects about to transform our planet, our civilization, our society— and our lives.

Competing for the World's Largest Infrastructure Project: Over 100 Million Jobs at Stake

Both Elon Musk and Daryl Oster are working on what will likely be the next generation of transportation where specially designed cars are placed into sealed tubes and shot, much like rockets, to their destination.

While high-speed trains are breaking the 300 mph speed barrier, tube transportation has the potential to make speeds of 4,000 mph a common everyday occurrence. As Daryl Oster likes to call it, "space travel on earth."

Even though tube travel like this will beat every other form of transportation in terms of speed, power consumption, pollution, and safety, the big missing element is its infrastructure, a tube network envisioned to combine well over 100,000 miles of connected links.

While many look at this and see the lack of infrastructure as a huge obstacle, at this point in time it is just the opposite, the biggest opportunity ever.

Constructing the tube network will be the biggest infrastructure project the earth has ever seen, with a projected 50-year build-out employing in excess of 100 million people along the way.

In addition to these impressive projections, there's far more at stake than just jobs and superfast transportation.

Transportation Trends

According to Richard Florida, author of the best seller *Rise of the Creative Class*, average transportation speed in the U.S. in 1850 was 4 mph. As more cars and trains came into use, by 1900 speeds had doubled to 8 mph. Driven by the Henry Ford car era and an emerging airline industry, by 1950 the pace of travel tripled to 24 mph.

With airline travel becoming far more common, by 2000 the average was boosted all the way to 75 mph. Following this trend line, the logical next iteration of travel will boost averages to 225 mph or more.

So what is the breakthrough in transportation that will move us to a whole new level of speed and efficiency? Many are beginning to think tube transportation is the logical next step.

Early History—The Vactrain

For nearly a century, this form of future travel was being referred to as the "vactrain concept." Russian professor Boris Weinberg proposed a "vactrain" concept in 1914 in his book *Motion Without Friction*. He also built an earlier model at Tomsk University in 1909.

The vactrain concept was also being studied in 1910 by American aerospace pioneer, Robert Goddard, who created a detailed prototype with a university student. His train was designed to travel from Boston to New York in twelve minutes, averaging 1,000 mph.

The train plans were found only after Goddard's death in 1945 and shortly thereafter his wife filed for the patents. Vactrains later made headlines during the 1970s when a leading advocate, Robert M. Salter of RAND, published a series of elaborate engineering articles in 1972 and again in 1978.

Vactrains also appeared in science fiction novels, including Arthur C. Clarke's *Rescue Party* (1946), Ray Bradbury's *Fahrenheit 451* (1950), and Robert A. Heinlein's *Friday* (1982).

Launching the Tube Transportation Era

Daryl Oster's epiphany moment happened back in the 1980s in a mechanical engineering class in college when he was calculating the drag coefficient on various shaped objects in a wind tunnel and made a mistake with air-density. On a lark he dropped the air-density to zero and epiphany struck: It suddenly occurred to him how beneficial it would be to travel in a vacuum.

Over the following decades, designing ET3's vacuum tunnels and maglev tracks became an obsession for Oster, as he oriented his work and research around the massive benefits of frictionless travel, forming the original company in 1997.

In 2012 Oster formed the ET3 Global Alliance to serve as a licensing consortium to create an open opportunity for key companies and individuals around the world to participate. The Alliance allows for easy pooling of technology and intellectual property along with equally simple licensing of the technology.

There are three differentiating features in the ET3 design. First it's built around a narrow tube diameter to reduce weight and maximize vacuum efficiency. Narrower tubes, only 5' in diameter, mean less vacuum pumping, lighter pylons and bridge supports for elevated segments and less drilling when going underground or through mountain ranges.

Second, Oster's capsules are relatively small, designed around the dimensions of a midsize car, 4'3" high and 16'2" long. Small capsule sizes mean lower costs for things like the yttrium barium copper oxide ceramics on board to maintain superconductivity, and less cost for life-support and entertainment systems.

Each capsule will have room for up to six seats for passengers or three pallets for cargo. The maximum weight, including passengers, baggage, and cargo: 1,212 pounds. Minimal sized capsules means less stresses and lower costs throughout the entire system, translating into massive cost savings, operating at one-tenth that of high speed rail or a quarter of the cost of cars on a freeway.

The third differentiator is the use of high-temperature superconducting maglev, which ET3 licensee Yaoping Zhang pioneered at China's Southwest Jiaotong University. The technology uses liquid nitrogen rather than liquid helium as a coolant, which lets the system run somewhere between 63 and 77 Kelvin—minus-321 to minus-346 Fahrenheit—the zone in which nitrogen neither boils nor freezes solid. Traditional maglev runs on helium which is much more expensive. The capsules will have the superconductor material onboard.

Hyperloop's Background

Elon Musk first mentioned he was thinking about a concept for a "fifth mode of transport", calling it the Hyperloop, in July 2012 at a PandoDaily event in Santa Monica, California.

He described several characteristics of what he wanted in a hypothetical high-speed transportation system: immunity to weather, cars that never experience crashes, an average speed twice that of a typical jet aircraft, low power requirements, and the ability to store energy for 24-hours of operation. He estimated at the time that the cost of the Castro Valley-Sylmar Hyperloop would be about US$6 billion.

Musk has likened the Hyperloop to a "cross between a Concorde and a rail gun and an air hockey table," while noting it has no need for rails. He also noted it could work either above or below ground. From late-2012 until August 2013, an informal group of engineers at both Tesla and SpaceX worked on the conceptual foundation and modeling of Hyperloop, allocating some full-time effort to it toward the end.

The tubes would maintain a vacuum-pressure equivalent to an altitude of 150,000 feet. This is very thin air, but still 1,000 times denser than ET3's proposed vacuum, and therefore easier to manage leakage and entry and exit of capsules through airlocks. But even that tiny amount of air is enough to dramatically increase demands on the capsules, which include a vacuum engine powered by a 436-hoursepower motor.

A high-level alpha design for the system was published on August 12, 2013, in a whitepaper posted to the Tesla and SpaceX blogs. Musk also invited feedback as an open source design project to "see if people could find ways to improve it."

The following day he announced a plan to construct a demonstration of the concept.

The World's Largest Infrastructure Project

Once a technology sets a new standard, no one wants to be left out. Like any radically new technology, tubetransport starts with a level playing field. A pilot project will lead to the first city-to-city project, and once successful, a rush-to-be-next will ensue. A global consortium will be assembled to map out plans for international trunk lines, and individual countries will begin thinking through feeder line strategies to connect to the cross-continent central system.

Within a few years the vision will morph into a tangible reality, and like road-builders in the past, schools and training systems will crop up around the world, and construction will begin. Even before the main trunk lines are complete, an entire network of feeder lines will begin to crop up both for bragging rights and to help countries gain a better grasp on this new form of transportation.

It's important to understand that even with the most optimistic scenarios, decades will be needed for the complete build out. All told, this new transportation system will cost well over $1 trillion to construct, creating over 100 million jobs over the next fifty years.

The Value of a Super-Connected Society

Besides cutting pollution and dramatically lowering our carbon footprint, faster and cheaper transportation will lead to a far more connected world. The number of people crossing country boarders each year will grow from millions to billions, and conducting business on seven continents simultaneously will become as common as our cross-cultural thinking.

A super-connected society is also a dependent and *inter-dependent* society. More than ever, people will learn to need and respect each other. That's not to say there won't be outliers who want to destroy much of what is being built, but the majority of people will shift their thinking from micro-neighborhoods to macro-neighborhood.

At the same time, unique talent will become more discoverable. Artisans and craftsmen will all be able to carve out their own niche. Serendipity will grow exceedingly long arms, and once-in-a-lifetime meetings and events will begin happening with far greater frequency. Along with increasing levels of both physical and digital awareness, the IQ of the entire planet will climb significantly.

It Pays for Itself

Back in 1972, RAND's Robert Salter wrote, "We no longer can afford to continue to pollute our skies with heat, chemicals and noise, nor to carve up our wilderness areas and arable land for new surface routes. Nor can we continue our extravagant waste of limited fossil fuels."

Today, the U.S. population is fifty percent larger; U.S. airline passenger miles have leapt by a factor of twenty we drive, collectively, 250% more miles in more than twice as many vehicles; and our atmosphere is laden with twenty-one percent more carbon dioxide.

Vacuum tube transport is not just a great idea; it's becoming a moral imperative. Ships and planes are polluting our oceans and skies faster than nature can clean it up. This is a solution that will not only solve all those problems; it will create over one hunderd million jobs along the way.

Most importantly, tube transport will pay for itself. If it were on this year's ballot, I would vote yes.

AS AWARENESS GROWS, COUNTIES WILL SOON FIND IT NECESSARY TO COMPETE FOR THEIR CITIZENS, SOMETHING THEY'VE ALWAYS TAKEN FOR GRANTED IN THE PAST.

LATE ADOPTERS TO THIS STRATEGY WILL QUICKLY FIND THEMSELVES LOSING THE TALENT POOLS NEEDED TO COMPETE IN THE GLOBAL MARKETPLACE.

AS TECHNOLOGICAL UNEMPLOYMENT GROWS, COUNTRIES WILL BE LOOKING FOR MEGA PROJECTS TO BOTH EMPLOY AND REEMPLOY OUR YOUNG PEOPLE BOTH NOW AND FOR GENERATIONS TO COME. EMERGING TECHNOLOGY AND AUTOMATION WILL MAKE THESE TYPES OF FUTURE MEGA PROJECTS BOTH AFFORDABLE AND TECHNICALLY DOABLE.

EPIPHANY Z—
OPTIMIZING OURSELVES

Human Life Calculations

What's the value of a human life?

For some of you this is a very disconcerting question because it attempts to put a dollar value on a person, something we value in far different ways. Yet this is exactly what governments and businesses find themselves doing on a daily basis. Every time an insurance company calculates premiums, or militaries plan budgets, or juries calculate liability awards, the value of human life is central to their decisions.

In fact, the value of people is a subconscious calculation we each make on a daily basis. Each of the following statements reflects a value judgment taking place in the back of our mind:

If I take this training my salary will go up.
When the mayor died, his estate was worth millions.
As a single mother raising 7 children, she left a tremendous legacy.

Much like adding an adjustment for inflation, cost of living increase, or paying a premium for a popular brand-name, we are constantly readjusting our sense of life's value in our mind. To some, the difference in value between a homeless person in Indonesia and the President of the United States may be well over $1 trillion. To others, the two individuals should be considered equal.

Seven global shifts are currently underway that are causing the underlying value of human life to move up the exponential growth curve, and along with the upward movement, a massive reassessment of corporate decision-making is about to begin.

Here is why this will become such a huge factor over the coming years.

Seven Reasons Why the Value of Human Life is Increasing Exponentially

To set the stage, there's a fascinating video clip of a debate about the value of human life between the late Nobel Laureate Milton Friedman and a young Michael Moore. In this exchange Moore objects to a decision that was made by Ford Motor Company in the 1970s, based on a typical cost-benefits analysis, to not spend $11 per car on changing the design of the Pinto gas tank to reduce the likelihood of gas-tank explosions. Friedman contended that Moore's complaint merely was over the low value of $200,000 per life lost, not over the principle that the value of human life has a finite upper limit.

Moore seemed to agree with that principle, but he objected to the idea that a Ford executive could casually decide the fate of Pinto buyers, and the value of avoiding a horrible death or injury from a burning

Pinto was as low as the company assumed in its formal risk analysis. Moore assumed that most Pinto owners would have gladly paid the extra $11 to fix the gas-tank problem.

Business decisions like this are far more complicated than Moore implies literally thousands of tradeoff decisions are made on the design of every vehicle, 100% safety can never be guaranteed, and cost savings is always a significant factor.

But—this sort of cost-benefits analysis looks dramatically different if the value of a human life were to shift from $200,000 to as high as $2 billion.

If you're wondering how much your life is worth today, you're not alone. Here are some of the factors that will drive "value-of-life" calculations like this through the roof in the future:

1.) Declining Birthrates
Global populations are growing increasingly fluid, and currently the United Kingdom is home to the most diverse immigrant community in the world, where 1 in 8 people are immigrants.

But the most significant change occurring is a global depopulation trend happening in most wealthy nations. Countries need to average 2.1 children per family to maintain an even population base.

According to the World Bank, here is what's happening in some of the most populated countries in the world:

1. **Korea (1.19)**
2. **Japan (1.43)**
3. **Thailand (1.56)**
4. **Russia (1.60)**
5. **China (1.69)**
6. **Brazil (1.81)**
7. **Chile (1.85)**

8. **U.K. (1.91)**
9. **U.S. (1.93)**
10. **Sweden (1.98)**
11. **France (2.00)**
12. **India (2.48)**

With the exception of India, each of these countries will experience shrinking populations in the future.

As birth rates decline, each child becomes more precious, and the value of each life rises.

2.) Improving Global Connectedness

We all have our own fan clubs—people who care about us and we care about them. In the past, our primary influence tended to be limited to our Dunbar Number, the 150-250 people we were closest to.

Today, with extensive social media connections, our fan clubs have grown not only to include the strong relationships found inside the Dunbar Number, but also weak and even tangential relationships with people all over the world. Over time, the value of our personal network in tomorrow's hyper-connected world will become markedly more quantifiable, and by extension, more valuable than the formulas we use to drive our connectedness metrics today.

3.) Improving Base of Skills

A skilled laborer is more valuable than an unskilled one, and a multi-skilled individual is even more treasured. Over time, our ability to accurately assess macro and micro skills will add to the growing body of evidence that the value of human life is indeed snowballing.

Counter to fatalist thinking that automation will cause large numbers of people to be unemployed, automation is simply readjusting our capabilities. By 2030, with the help of automation, the average person will be able to accomplish fifty to one hundred times more in their lifetime than their counterpart today.

4.) Increasing Life Expectancy

29,000 people in Japan celebrated their one-hundredth birthday in 2015.

Life expectancy is currently increasing by two years every decade, and there are no signs of that slowing down. Average lifespan around the world is now double what it was 200 years ago.

Most experts have concluded there really are no hard ceilings that will prevent us from living indefinitely.

In Great Britain, as example, the Office for National Statistics predicted in 2010 that nearly one-in-five people would live to see their 100th birthday. By 2030, that number could well exceed one-in-two (fifty percent).

As the producing/consuming years of human life grows, so does its overall value.

5.) Increasing Options for Creating Wealth

Rank and file executives within the banking community are outsiders to the emerging cryptocurrency movement with their thinking focused on "how they will fail." On the other side of the equation cryptocurrency entrepreneurs, some of the best and the brightest in the world, are continually asking, "How can we make this succeed?" In addition to physical wealth, we have created numerous ways to accrue less tangible forms of wealth such as owning property rights, digital assets, and intellectual property.

6.) Decreasing Poverty Rates

At the same time that global wealth is increasing, extreme poverty is dropping. We still have a long way to go to create what many believe to be an equitable distribution of wealth around the world, but the trend is moving in the right direction. Along with decreasing poverty comes increasing purchase power among even the poorest of the poor.

7.) Accelerating Sense of Preciousness in Children
Most families today are fine with only one or two children. Dropping from families with seven to ten kids fifty years ago to less than two today, the amount of time and attention dedicated to each child increases. From an investment standpoint, parents today are willing to pull out all the stops, viewing everything from better daycare, to better clothes, to better sport leagues as an investment in their children's future.

Long-Range Implications
These seven major trend lines, combined with dozens more, will be causing us to continually rethink how we value human life. As the value of people climbs into the stratosphere, it will have huge implications on everything from product liability cases, to life insurance, to the way we value ourselves.

> "People in the future will view themselves
> as being in a constant state of improvement."

This means over the coming decades we will become exponentially more fixable—trainable, repairable, improvable, and even *re-inventible*.

Our lives will never be about who we are today, but who we have the potential to become.

Capturing Real Human Intelligence

We all dream of an easier life.

Suppose we got into our car and it knew where we wanted to go, or we turned on a radio and it played the perfect music, or we pressed "call" on our phone and we would instantly be connected to the person we most wanted to talk to—all without our having to do anything other than take the initial action.

In reality, our days are filled with countless decisions, and the stress level of all these choices is growing steadily. We want to be in control, but control can be very taxing.

We tend to worry about computers that are smarter than we are, automating our skills and taking our jobs. But if computers become more human-like in their thinking, adding our own emotional values to everything we think is important, the heartless machine-only qualities of these technologies will disappear, moving computers away from the paradigm of human-replacer toward something more akin to human-enhancer.

Artificial intelligence only goes so far. But finding a way to capture pieces of real human intelligence can give us far more pertinent information. As an example, if someone conducts a typical search on a search engine, the connecting path between the search terms and the final destination is a very real piece of human intelligence.

Certainly not everyone will agree on the final site selected from a set of search terms. But that will improve over time. According to Futurist John Smart, in 1998 the average online search phrase consisted of two words. Today, the average search contains 5.2 words, trending towards something far longer, more akin to a natural question of eight to ten words. Over time, capturing millions of "paths" will yield a base of growing intelligence based on the cumulative thinking of everyone involved.

Refining the Relevance Algorithm

Gone are the days when computers only recognized keyboard inputs. Combining a variety of ambient information, such as the user's social graph, search and surfing history, dwell-times associated with photos and videos, purchase history, and even music preferences helps define who we are, providing the computer a far more precise approach to delivering what we are asking for. Computers can't read our minds, at least not yet. However, by paying close attention to all of our inputs and outputs, the information we consume and how we respond when we consume it, anticipatory computing can develop a very close approximation to the inner working of what our mind is telling us to do next.

Moving one step further, by capturing verbal conversations surrounding videos and photos, we can begin to automate the metadata and searchable tagging systems surrounding both physical and digital objects. Thinking even farther into the future, as users add sensory capturing devices to their clothing, phones, and skin (as in tattoos), we will begin to translate emotional data into a complex value system that can better determine our opinions without us having to personally make decisions.

We have gone from crude computational devices like the abacus and slide rule, to punch card readers, to memory storage and retrieval devices. By adding the complex communication capabilities of the Internet to our once isolated computers, we have begun an exponential climb towards developing a sort of cyber humanism.

Until now, computers have been trending along the bottom part of this transitional curve, going from cold, heartless machines to something more warm and fuzzy with human-like qualities. In the very near future, computers will be christened with a higher purpose than making us slaves of the Internet. Instead, they will be destined to make us more human, even ultra-human.

So the real difference in our future relationship with computers is not an "either-or" proposition that pits humans against computers, but rather a "both-and" option where we use computers to enhance our abilities and magnify our own personalities. Certainly someone who is prone to making bad decisions could theoretically make ten times as many bad decisions in a single day. But this will just be another step in the ongoing evolution of computers.

There is no magic formula for solving every problem. But anticipatory computing has the potential to ratchet up human capabilities and intelligence to make decisions and solve problems far faster than anything before.

Introducing Synaptical Currency Theory— Assigning Value to Brain Capital

What has been the hardest problem you've had to solve in your life? As I step you through this question, just focus on the ones where you actually found a solution.

For some of you, this may have included things like finding a job, finding a cure for a disease or medical problem, or dealing with major family issues. When you had to solve the problem, how much time, energy, and brainpower did you commit to coming up with an answer? Now consider how different your life would have been if it only required half as much effort—half as much stress, anxiety, and mental turmoil.

Assuming there is a physical limit to the number of times a synapse triggered signal can pass through the human brain in one day, the way we expend our "synaptical currency" becomes a crucial element in our personal success formula. So how do we go about assigning monetary value to the finite resource I'm calling "synaptical currency" that will eventually determine our value in society?

How Much is Brain Power Worth?

When it comes to finite resources, brainpower is an obvious one. Every one of us runs into natural barriers in our ability to consume and process information.

If we were able to place some sort of synapse monitor next to our heads that actually counted the number of signals passing through our neurotransmitters on a daily basis, a number I'm sure varies widely from person to person, we would begin to see quantifiable limits on our individual capabilities.

It's easy to speculate that those with higher limits are the brightest, but that's probably not the case. What's more important is there are limits, and how we expend any of our finite resources becomes a critical component of personal achievement.

Personal Accomplishments

What do *you* consider to be your greatest accomplishment in life? This question is similar to the previous one about the hardest problem you've had to solve. These sorts of questions force you to become introspective, viewing yourself from the inside out.

Most people have great difficulty with this question because what we've personally accomplished seems rather minor in comparison to what most would consider worthy of being called the "greatest." However, if you reframe the question around what you'd *like* to accomplish, and compare it to what you have already accomplished, you begin to see the gap between the two.

In the past, most people's greatest accomplishment was simply survival. The vast majority of their day was spent on finding food, water, shelter, and staying one step ahead of predators, extreme weather, or whatever it was that could kill them.

In that era, very little time was dedicated to the sorts of things we would consider significant accomplishments today. People were too busy working at staying alive to pursue less immediately practical matters. But as society progressed and became more systematized, far less time was being dedicated to survival, allowing more energy to be directed towards more esoteric endeavors like what we want to accomplish.

Twice as Much in Half the Time

Do all of our synaptical expenditures have value, perhaps internally, but not externally? We spend our synaptical currency both when we learn and when we work. We also spend it on fun and entertainment. We spend it every time we make a decision, and even while we are sleeping.

Most of us hold little regard for how we often squander this limited resource except when it comes to work. The synaptical currency we sell to others, in the form of work, typically has to be accompanied by a paycheck commensurate to the value of our work.

As we add unique and different forms of automation to our lives, the amount of synaptical currency we dedicate to accomplishing individual tasks begins to decline, and our output becomes less brain intensive.

With the automation of work, the expenditure of synaptical currency per accomplishment will decline, but so should our expenditures for learning. If we can accomplish twice as much in half the time, we should also be capable of learning at a comparably quicker pace.

Few people have attempted to quantify synapse firings in the brain, but according to neuroscientist Astra Bryant, a rough number for neural signal transmissions in the average brain ranges from 86 billion to 17.2 trillion actions per second.

Certainly I don't claim to be a brain expert even though I've been using mine for well over 10,000 hours. I'm not even an expert on using my own brain since there are countless ways I could be using it more efficiently. If we take Gladwell's 10,000-Hour Rule and subdivide it into synaptical transmissions, we will end up with a very large number. But the number is not infinite.

Similarly, if we assign a dollar value to our synaptical currency, yes it would seem infinitesimally tiny in comparison to the number of neurotransmissions needed to equal a single dollar. But again, it's not zero. Over the coming decades we will be creating a world that works with exponentially greater precision than how we operate today.

Accomplishments that can be completed through a fraction of today's synaptical expenditure will give rise to far greater accomplishments in the future. Admittedly, my theory of synaptical currency is far from complete, but I do firmly believe "synaptical currency" eventually will be used to determine our value in society.

The Perfection Quandary

Ironic as it may sound, perfection is an imperfect concept.

Each of us has been born and raised with all of the foibles and limitations of being human. A typical day involves forgetting where we put our keys, stubbing our toe, getting angry at the wrong person, and dropping a plate full of food. And those are just the little things.

We are indeed intelligent beings, but for all our limitations, the intelligence we possess hardly seems enough.

Even after a full night sleep we still wake up tired, we crave food that isn't good for us, and our pets end up being a poor substitute for the kids we didn't have.

Most importantly, we have a never-ending need for social interaction. The proliferation of social networks has given many the illusion of being surrounded by those who love them, but the reality is just the opposite. A 2012 study in *The Atlantic* concluded that only one in four Americans have someone with whom they can discuss important matters, compared with one in ten 30 years ago.

Similarly, a 2013 survey conducted by *Lifeboat* shows that the average American only has one real friend, drawing the conclusion that we are in a friendship crisis. This prompted *The Guardian* newspaper to declare we are entering the "age of loneliness."

The opposite of loneliness is not togetherness, it's intimacy. We need to be needed, and that's where perfection comes in.

Our human need is what creates our economy. Without needs, we have no economy.

It's easy to imagine a perfect person as being self-fulfilled. The more flawless our lives become, the more self-sufficient, self-reliant, self-absorbed, and isolated we become. But our need for control works just the opposite, leaving us in control of a universe of one.

Nothing exciting ever happens in a vacuum. Well, it does, but the vacuum doesn't care. We need someone who cares.

That's where artificial or machine intelligence comes into play. The improvements we seek with this technology are rarely going to be the improvements we need.

Our striving to make a better world is a superficial goal, much like winning the lottery, buying expensive jewelry, or eating three pounds of chocolate. The instant high we experience from filling our immediate gratification only sets us up for a second stage of emptiness and the crashing realization that once we have it all, it's never enough.

Our Need to be Needed

Most of us today are incredibly lethargic, sitting 9.3 hours a day. Sitting has become the smoking of the Millennial generation.

But from our personal command center, wherever it is that we may be sitting, we're able to control most of what we want in our lives, right?

So what is it we want to control? On-demand entertainment, on-demand answers, food, healthcare, sex, transportation, news, or something else.

Setting perfection as our goal, we need to begin with defining perfection. Does perfection mean we've optimized our efficiency, our purpose, our income, our accomplishments, our relationships, our happiness, or something else?

The balancing act of life was never intended for someone to win in every category, and even if that were possible, without needs we'd somehow become devoid of purpose.

This entire discussion has left me in a bit of a quandary, better put, a perfection quandary. Why is there an exception to every rule? Perhaps we need to solve the laws of unintended consequences?

Try as we may, we seem destined to struggle.

For this reason, I've concluded that broad forms of AI will not live up to their expectations, and the singularity will not unleash the utopia many are predicting. But as with all mysteries of science, we will never really know for sure until we reach the other side.

When Death Becomes Optional

The year is 2040. You have just celebrated your 80th birthday and you have some tough decisions ahead. You can either keep repairing your current body or move into a new one. The growing of "blank" bodies has become all the rage, and by using your own genetic material, body farmers can even recreate your own face at age twenty. In just twenty years, this is an industry that has moved from the equivalent of Frankenstein's laboratory to the new celebrity craze, with controversy following it every step of the way.

The combination of a few high profile "accidents" along the way, coupled with those in the religious community who claim body farmers are playing God, and asking "where does our soul reside?" has given it thousands of top media headlines around the world. Every person on the planet has a different opinion about this moral dilemma, or whether it's safe or dangerous, or whether we should just get better at repairing our existing bodies. As medical advances continue, and we devise an entirely new range of health-enhancing options, I propose we set a new standard, raising the bar to the highest possible level. I propose we put an end to human death.

In the coming years we will find ways to fix human aging, cure diseases, find solutions for deviant behaviors, and even rebuild people after an accident.

In short, no person should ever need to die… EVER! Is that our goal? Is that the direction we are headed in? If that's not our goal, then we will need to hear the pro-death arguments, and why people should die when they don't have to.

Is the goal of the medical community to improve health, or to completely eradicate health problems? In Star Trek terms, what is the prime directive for the health profession? In the years ahead we will have an unbelievable number of tough choices to make.

Dealing with the Issues

It is an interesting exercise to look closely at each cause of death and think through not only how to reduce its influence, but how to eliminate it completely.

For trauma cases, the medical profession will be grappling with the ethics of making simple repairs versus making better, longer lasting humans. These questions no longer belong to the realm of science fiction. In a few years, we will be able to replace our frail parts with parts that are made of superior materials.

Recent medical accomplishments include everything from re-growing bladders and throats to using 3D printers to "print" new bones and arteries. Entire body replication can't be far behind. We may have to ask if the path we've taken to this point in medical history has come too quickly. Have we really absorbed the impact of what these changes will mean for society?

We have wrestled with certain ethical questions, sometimes for centuries. Some issues that are clearly outrageous in our in minds today, such as slavery, had to be abolished at the point of a gun.

Now we are about to be presented with the question: Will we become something else? Do we have a right to live indefinitely, or is it a privilege? Life, death and years of painful adjustment, all can be avoided. Families will be spared the emotional turmoil of deciding life and death for the brain damaged. The horribly burned can re-grow and shed their charred skin. In a nation where health care is afforded to some, but not all, be prepared to take on such questions as to who lives forever? And more importantly, who will pay for it?

The prisons and jails are strained. Science may even address the morally and mentally challenged. Is it realistic to think we can fix the underlying behavioral issues, turn criminals into productive, well-intentioned people? Should they be allowed to live forever, too?

The End of the End

Let's play a game of "what if".

What if we could turn our attention from fixing problems of the past to pondering a new kind of future? What if we could advance civilization to the point where past problems begin to go away completely? I propose something quite radical in today's world—that we declare war on human death.

Advances for cures and minor diseases have moved glacially since the start of time. From Leeuwenhoek's invention of the microscope in the late 1600s to Louis Pasteur's discovery of germs in the 1800s, advancements obviously took centuries.

Today, breakthroughs are arriving at greater speed and accelerating to the point where barriers to near-term immortality are falling daily.

The most dramatic advancements have been seen in the quickening speed of communication, and the spread of knowledge across the Internet. Breakthroughs are commonplace.

Online, science accomplishments are building on other accomplishments as never before. Virtual collaboration has led to global teaming. As information speeds to all corners of the world, the approaches to solving some of our most perplexing problems have multiplied. Warring with human death requires a far different mindset. The current trend of paying to live is not the model for defeating death. Here are a few examples of the hurdles that would have to fall:

Redefining Aging—Humans will need to be re-engineered to stop the aging process at around twenty-five years of age when bodies are in their peak condition. Humans could live indefinitely at the peak of health.

Keen Minds—Human mental condition deteriorates as brain cell death takes its toll. Scientists have discovered a replicating switch in individual optic and brain cells that for some reason is turned off soon after birth. Throwing the switch back on is in the cards. Epilepsy, blindness and dementia will be relegated to the past.

Accidents—Accidents are inevitable. Is it reasonable to think that all can survive accidents?

Mangled Bodies—Should we rebuild bodies that become hopelessly—by today's standards—damaged? Falling into a wood chipper or stepping in front of a steamroller need not be fatal. Is it possible to reassemble the body and reinstall memories?

Terminal Illness—Can we put an end to viral and bacterial diseases that ravage our bodies from within? Will pathogens become harmless artifacts? With the advent of blood-roaming nanobots, the future will be bleak for the little animals that ravage bodies. Cancer will be forgotten.

Criminal Minds—Are evil people really worth saving, even if they can be reformed? Science has barely scratched the surface when it comes to understanding the brain. Is it possible to live in a placid society that has forgotten the fear of crime by sociopaths and others lumped under the rubric of the criminally insane?

End of War—We will always have conflict. So, how do we resolve such disputes among societies that no longer know or appreciate the meaning of death? If war becomes convenient and non-traumatizing, can it be loathed as it is today?

Death as a Motivator—Nothing motivates like the anticipation of a deadline. With no fear of death, what will become of our humanity? With no need to appreciate death, what becomes of the forces that drove generations to confront such weighty matters as meaning and challenge? Will our imperatives lose their potency; even disappear? Will we pursue achievement in the absence of these possibly extinct drives?

Change is coming at lightning speed.

We don't have the luxury of mulling these matters of mortality for decades. If death is no longer viewed as inevitable, our attitude towards life will shift dramatically. Though the challenges seem overwhelming, I believe this is the time to establish a long-term directive, a prime directive for all—the end of human death. Even if it proves not to be achievable, shouldn't it still be our goal?

OUR GENERATION MAY WELL BE THE
LAST WHERE DEATH IS NOT OPTIONAL.

EPIPHANY Z—
THE ULTIMATE EPIPHANY

N o future offers a completely blank slate—as I've shown, all futures build upon the past. And those who will thrive best in the future will be those who understand and can draw upon and learn from past information.

With that in mind, I'd also point out the value of thinking of the future not as a blank slate, but as a vast and undeveloped territory. A territory the size of our entire planet.

New Planet Scenario

I often think about what it would be like to colonize a new planet and start a new civilization from scratch. Starting with a clean slate, and knowing what things work well and not so well on Earth, how could we construct a significantly better society? As with every society, it begins with creating a series of new systems, and these systems are all formed around rules.

Rules create order. They create the inter-relational fabric of society around which all of our actions are woven.

"Much like colonizing a new planet, we are just now coming to grips with the need for rules and order in the emerging digital information age."

Eight Critical Skills for the Future

Although equal in importance to social systems, we currently have very few rules for how to live our lives in a fully immersive world where explosive amounts of information are flowing to us and around us on a second by second basis. Since each of us interacts with this information differently, it is up to us to master the "new rules of engagement."

With that in mind, here are eight skills I see as being critically important in our future:

1.) Communication Management—How much is too much? According to Nielsen, teenagers in the U.S. sent and received an average of 3,276 texts per month in the last quarter of 2010. A Pew Research Center study from 2010 reported that more than four out of five teens with cellphones sleep with the phone on or near the bed, sometimes falling asleep with it in their hands in the middle of a conversation.

Pew's Amanda Lenhart, a senior research specialist, said, "many expressed reluctance to ever turn their phones off." It's getting to the point where hospitals are starting to see young patients who come in exhausted from being "on call" or semi-alert all night as they wait for their phones to vibrate or ring with a text. Communication is an essential ingredient in all of our lives, but too much or too little can have devastating effects.

With new communication channels springing to life in games, social media, and smartphone apps on a regular basis, people suffer great anxiety over not keeping up with their friends and family. And when they turn things off, they suffer even greater anxiety over feeling left out. Effective ways of managing our communications is a critical skill not currently being taught in school.

2.) Reputation Management—Our reputations are no longer something that builds up around us that we have little or no control over. With highly personal online content being generated about us from many different sources, it is now up to us to exercise control over what people are saying, the images of us that appear online, videos we're in, bylines of our work, and virtually every other indicator of who we are and what we stand for.

About fifty-seven percent of adult Internet users in the United States said they have entered their name into a search engine to assess their digital reputation, according to a 2010 Pew Research Center study, "Reputation Management and Social Media." That's a ten percent increase since 2006, when only forty-seven percent of adult Internet users said they had queried their own name. TAs the study notes, "reputation management has now become a defining feature of online life."

The study also found that young adults are more apt to "restrict what they share" and manage their online reputations more closely than older Internet users. This is "contrary to the popular perception that younger users embrace a laissez-faire attitude about their online reputations. Clearly this is another critical skilln schools have yet to come to grips with.

3.) Privacy Management—Privacy and transparency live at opposite ends of the same social spectrum. Pew also studied online privacy study and found that social networkers ages 18 to 29 were the most likely to limit their profile privacy settings. The percentage that did so was seventy-one percent, compared with just fifty-five percent of the fifty to sixty year-old bracket. Altogether, about two-thirds of all social networkers who were surveyed said they've tightened security settings.

People derive significant benefits from sharing their personal details as they take advantage of relevant and useful services online. However, once collected, businesses often exploit and monetize personal information, leaving people exposed and placing their information in predatory danger. Yes, protecting and enforcing privacy is an added burden for business, but a lack of privacy creates risk for users and reduces trust. Trust plays a key role in innovation.

The free flow of personal information that respects privacy can fuel and cultivate innovation. Optimizing the risks and rewards across the stakeholders may lead to new forms of innovation and the release of new economic value. The big challenge ahead will be to establish legal frameworks that foster innovation and facilitate information sharing across jurisdictions in global business environments. Understanding both sides of this equation will be a critical skill for future generations.

4.) Information Management—In 2008, Roger Bohn and James Short, researchers at the University of California in San Diego, did a study to determine the amount of information people have entering their brains on a daily basis.

In rough terms, 41% of the information is received from watching television; 27%—computers; 18%—radio; 9%—print media; 6%—telephone conversations; 4%—recorded music; and smaller amounts from movies, games, and other information sources.

The average American spends 11.8 hours consuming information each day. People today are being exposed to far more information than ever in the past.

How can we manage all this information better? How can we be smarter about the information we consume and the sources we're getting it from? Our ability to effectively manage our personal information inputs and outputs will greatly determine our ability to compete in the global talent marketplaces of the future.

5.) Opportunity Management—The average person who turns thirty years old in the U.S. today has worked eleven different jobs. I'm predicting that in just ten years, the average person who turns thirty will have worked on 200-300 different projects. Short work project employment will replace long-term employment for many.

Business is becoming very fluid in how it operates, and the driving force behind this liquefaction is a digital network that connects buyers with sellers faster and more efficiently than ever in the past. Opportunities are springing to life all around us. Having an ability to find, select, and capitalize on opportunities will be a critical ingredient in how successful people run their lives in the future.

6.) Technology Management—New tools are entering our lives on a minute-by-minute basis. What should we be paying attention to, and what can we dismiss?

Our choice of technology defines who we are and our ability to function in an increasingly technology-dependent world. The tech-selection process has been largely relegated to tech insiders and key influencers with product manufacturers often playing a key role.

However, technology management goes far beyond hardware and software purchases. Both tend to evolve over time and the functionality is shifting on a daily basis with new apps giving us tools we never dreamed possible before. Our relationship with our personal technology will continue to be an ongoing challenge and improving skills in this area will be highly advantageous.

7.) Relationship Management—In a world immersed in social technology, we know lots of people, but what kind of relationship do we have with them? How do we qualify the value of those relationships? As the size of a person's social network increases, it becomes more difficult for someone to have meaningful conversations with each person in their network. Different rules apply to those we have strong ties with versus those who we maintain only a weak relationship with.

The way relationships are managed in the digital age is changing, especially when it comes to marriage. Contrary to the way a traditionalist would have it, for most college-educated couples, living together is like a warm-up run before the marital marathon. Couples work out a few of the kinks and do a bit of house-training and eventually get married and have kids. Those without a college degree tend to do it the other way around—move in together, have kids and then aim for the altar.

Our understanding of the shifting nature of relationships will be one of our most critical skills to manage in the future.

8.) Legacy Management—How will future generation remember you? How will they perceive your successes and failures, your accomplishments and misguided efforts, and your generosity and perseverance? While many still view inheritance as the primary way to leave a legacy, people now have the ability to manage the information trail they leave behind. In fact, they can very easily communicate with their own descendants who have not even been born yet.

The body of work we leave behind will become increasingly easy to preserve. So if we chose to let future generations know who we are and why we set out to achieve the things we did, we can do that today with photos, videos, and online documents. Future generations may even have the ability to preserve the essence of their personality and make interactive avatars that can speak directly to the questions and issues future generations will ask. As all of us age, the notion of leaving a legacy becomes critically important, and furthering our skills in this area will serve us well.

In addition to what I view as the eight "new" skills are two traditional skills that need to be radically updated to mesh with the needs of today's world:

- **Time Management**
- **Money Management**
  Time management classes of the past are a poor fit for the incessant pace and demand of living digital, and money management takes on an entirely new dimension with the any-time any-place tools at our disposal.

Fortunately, the new skills, and specifically our ability to acquire and master them, will be accompanied by a dramatic and vast increase in our capabilities.

The Laws of Exponential Capabilities

When people like Google CEO, Larry Page, Virgin's Richard Branson, and X-Prize Foundation CEO, Peter Diamandis, talk about us entering into a period of abundance, there has been a natural tendency to assume we'll be also entering into a life of leisure. People won't have to work as hard and we will all have more time for travel, vacations, and play.

We are entering into a world where driverless vehicles will eliminate millions of driving positions; robotic systems will work relentlessly day and night eliminating millions of manufacturing, welding, painting, and assembly positions; and things that seemed impossible to automate in the past will have computers and machines replacing people's jobs.

With these types of automation and AI replacing human involvement, the discussion has focused on solutions like shared jobs, micro employment, and guaranteed income. While those may be options, there's also great danger in preparing for "slacker lifestyles" where people feel less significant, less certain about their future, and less connected to the value they have to offer. As a society we risk becoming soft and lazy.

There is great value in the human struggle, and when we fail to be challenged, our best-laid plans tend to fall apart at the seams. Today, the amount of time it takes to build ships and skyscrapers, create massive data storage centers for all our growing volumes of information, or produce global wireless networks for all our devices has dropped significantly. Along with each of these drops is a parallel increase in our capabilities and our expectations.

Laws of Exponential Capabilities

LAW #1: With automation, every exponential decrease in effort creates an equal and opposite exponential increase in capabilities.

LAW #2: As today's significant accomplishments become more common, mega-accomplishments will take their place.

LAW #3: As we raise the bar for our achievements, we also reset the norm for our expectations.

LAW #1—With automation, every exponential decrease in effort creates an equal and opposite exponential increase in capabilities.

When it takes less effort to do something, we do more things. This has been proven time and again throughout the centuries. To illustrate this point, here are three industries that have radically changed humanity over the past centuries—Transportation, Photography, and Media.

1.) Transportation: Thinking in terms of our travel capabilities, if we use the average transportation speeds in Richard Florida's *Great Reset*, we can extrapolate an exponential growth in the number of miles the average person will travel over their lifetime.

- **1850**—Average speed 4 mph—Traveling 4 miles per day X 50 year life expectancy = 73,000 miles.
- **1900**—Average speed 8 mph—Traveling 8 miles per day X 60 year life expectancy = 175,200 miles.
- **1950**—Average speed 24 mph—Traveling 24 miles per day X 70 year life expectancy = 613,200 miles.
- **2000**—Average speed 75 mph—Traveling 75 miles per day X 80 year life expectancy = 2,190,000 miles.
- **2050**—Average speed 225-250 mph (projected)—Traveling 225 miles per day X 90 year life expectancy = 7,391,250 miles.

We have transitioned from slow and difficult forms of transportation to fast and painless. Going from 73,000 to 7.3 million miles in two centuries is a 100X increase in human mobility.

2.) Photography: The famous photograph titled, "*View from the Window at Le Gras*" by Nicéphore Niépce in 1826, was one of the first photos ever taken, and is the oldest surviving one. Photography started as a slow and arduous process in the 1800s requiring exacting precision and lots of time. With the introduction of cheaper and better cameras, film, and processing, the number of photos taken began working its way up the exponential growth curve.

It wasn't until recently, with the birth of digital cameras in our phones and free storage, the number of photos per day really took off. Currently, there are roughly 350 million photos a day loaded onto Facebook. If we assume the pictures loaded onto Facebook only represent a small fraction of the total, say 10%, that would mean we are taking 3.5 billion photos every day, or 1.3 trillion per year. As amazing as that sounds, that's probably a very low number.

3.) Media: Before the time of Gutenberg's printing press, our information sources were limited to person-to-person conversations and a tiny number of hand written scrolls and manuscripts. People who lived during the Middle Ages spent very little time consuming information simply because it wasn't accessible.

By 1600, India's Mughal Emperor, Akbar the Great, had accumulated a personal library of over 24,000 books. By comparison, in 1815, Thomas Jefferson had acquired the largest personal collection of books in the United States, totaling 6,487 volumes. Both of these numbers are in stark contrast to the millions of title available today on Amazon. But when it comes to media, we consume far more than just books.

On a global level, a 2012 study showed that people on average spend 10 hours 39 minutes per day consuming information. This was broken into 260 minutes on the Internet, 150 minutes watching television, 77 minutes mobile Internet, 71 minutes listening to the radio, 43 minutes playing games, and 38 minutes reading print media.

In countries like the U.S., Korea, and Japan, the numbers are considerably higher—over 12 hours per day—and China is now working overtime to reign in a growing problem with people becoming addicted to the Internet. As a result, a number of anti-addiction treatment centers have cropped up to deal with the problem.

LAW #2—As today's significant accomplishments become more common, mega-accomplishments will take their place.

It is no longer reasonable to assume the same mega-project that challenged us in the past will be the same size and scale of the mega projects needed to challenge us in the future. Living in a world where our ever-expanding use of automation and AI is reducing the human contribution in nearly every achievement, we are also witnessing a dilution in the value of past benchmarks. For this reason, new generations of mega accomplishments are beginning to surface.

One example of a next generation mega project is the Elon Musk—Daryl Oster proposed transportation system where specially designed capsules are placed into sealed vacuum tubes and shot, much like rockets, to their destination. While high-speed trains are breaking the 300 mph speed barrier, tube transportation has the potential of reaching speeds of 4,000 mph, turning it into a form of "space travel on earth."

Even though tube travel like this will beat every other form of transportation in terms of speed, power consumption, pollution, and safety, the big missing element is its infrastructure, a tube network envisioned to combine well over 100,000 miles of connected links. While many look at this and see the lack of infrastructure as a huge obstacle, it is just the opposite, one of the biggest opportunities ever. Constructing the tube network has the potential of becoming the largest infrastructure project the earth has ever seen, with a projected fifty year build-out employing hundreds of millions people along the way.

LAW #3—As we raise the bar for our achievements, we also reset the norm for our expectations.

When Pixar released the first *Toy Story* in 1995, it was the first feature film to be produced entirely with computer animation. Naturally it looked a little rough around the edges compared to the new stuff, but it represented a massive breakthrough in the way animated films were produced.

Fifteen years later, in 2010, when *Toy Story 3* was released, the Pixar team raised the bar considerably on the quality and detail of the animation. It didn't take them less time to produce, but instead they dedicated tremendous effort to raising the quality standard.

This raising of standards in quality, value, and usability can be seen all around us:

1. **Printing**—From large machine presses to photo-quality images at our desktop within seconds.

2. **Music**—From makeshift recordings inside seedy studios to producing symphony quality recordings without ever leaving our computer.

3. **Magazines and Newspapers**—We can now subscribe to any magazine or newspaper on the planet and have it instantly sent to our computer.

4. **Highways**—From dirt roads, to gravel roads, to asphalt roads, to concrete Interstates.

5. **Telecom**—From wired phones with cranks on the side, to wireless everything.

6. **Water Systems**—From aqueducts, to wells, to running water everywhere.

7. **Food Supplies**—From crude little storefronts and farmers markets to the super-grocery stores we have today.

8. **Emergency Services**—From makeshift fire brigades and primitive doctors to highly sophisticated fire departments, rescue teams, hospitals, and medical services.

Mega-Projects of the Future

Whether it's building the Great Pyramids in Egypt, erecting the Great Wall of China, or sending someone to the moon, crazy-big projects have a way of defining our humanity and raising the bar for future generations. As our capabilities improve, we simply need to set our sights higher and aim for the stars…. Literally!

A few mega-projects of the future may be:

1. **Recreating Infrastructure**—Virtually every one of our current infrastructures is in need of total overhaul to meet the needs of future generations. This includes rethinking highways, mass transit, telecom, postal systems, water supplies, food supplies, and more.

2. **Space Industries**—Whether it's space tourism, mining asteroids, space-based power stations, or colonizing other planets, space industries represent an endless challenge for humanity.

3. **Controlling the Weather**—We continually find ourselves the victims of forces of nature and have an obligation to mitigate the damage of hurricanes, tornadoes, massive hailstorms, and more.

4. **Reaching the Center of the Earth**—We currently know very little about the center of the earth, yet we continually fall victim to earthquakes, volcanoes, and other internal forces we don't yet understand. Once again, we have an obligation to know more.

5. **Controlling Gravity**—The single greatest force of nature is gravity, yet we know very little about it. We not only need to understand gravity, but also need to learn how to control it.

6. **Viewing the Past**—How can we create a technology capable of replaying an unrecorded event that happened decades earlier in actual-size, in holographic form?

7. **Traveling at the Speed of Light**—The all-time human speed record—24,791 mph—was set by Apollo 10 in 1969, and we have a long way to go if we ever intend to travel to other planets, much less other stars.

8. **Inexhaustible Power Supplies**—Too much of the world's economy is dependent upon a rather fragile global power system. The opportunities here are endless.

At this point, the "Laws of Exponential Capabilities" are components of a working theory that I will refine over time.

Those refinements include examining the negative consequences of exponentialism.

Naturally there are a few downsides to our expanded capabilities. Addictions can become exponentially more addictive. Dangerous people can become exponentially more dangerous. And global conflicts have the potential of becoming exponentially more disastrous.

With all of our increased capabilities, perhaps the one we are lacking the most is our ability to anticipate problems.

"We will all be spending the rest of our lives in the future, so each of us has a vested interest in understanding it better, both the positive and the negative. The better we understand both poles, the more fully equipped we are to steer our world towards a brighter, rather than darker future."

Inspiring a Better World Ahead

Having been born and raised on a small rural farm in South Dakota, I grew up with a very narrow perspective of the rest of the world.

With only two TV channels and three radio stations to pick from, our broadcast news options were very limited.

As a teenager, watching the nightly newscasts on television, I was thoroughly amazed at all of the things happening around the world, and yet none of them were happening near me.

I truly felt like I was living in a bubble, far away from all the excitement.

But I wasn't alone. People everywhere were still getting used to the new technology, and limited TV and radio access wasn't just a South Dakota issue.

For the most part, I didn't know what I was missing. Inside my bubble were all the families and neighbors I hung out with. Much like me, they didn't know what they were missing.

In understanding "bubble cultures," it's important to note there are micro-bubbles like the farm community I was raised in, and macro-bubbles that affect entire countries, planets, or civilizations as a whole.

Few people realize humanity today is being confined to a macro-bubble. Our limited grasp of today's technology, coupled with our limited understanding of the world, and just the limitations of being human, blind us from seeing our true potential.

In short, we're living our lives as bubble people, limiting our view of the world to what we know, what we can prove, and what "the experts" say is possible.

But the bubble we're in is not permanently confining or unbreakable. Over the past few centuries we have indeed been stretching the size and shape of our bubble, but even though it's far bigger today, we still have a long ways to go to see what's on the outside.

So is there an "outside" to our bubble?

The short answer is yes. In fact, the most exciting areas of the future will happen outside our current bubble. For this reason, I'd like to take you on a short journey to the other side of the bubble and an expansive view of human existence in the years ahead.

Eight Dimensions of Our Destiny—Pushing the Envelope of Human Existence

I've talked about how technological unemployment is a double-edged sword. On one hand people's jobs are being automated out of existence, but at the same time, we're freeing up human capital.

It's rather preposterous to think we're somehow going to run out of work in the world, but having jobs aligned with the work to be done is another matter entirely.

You've just seen the early stages of my "Laws of Exponential Capabilities," where the technologies being developed will give us exponentially greater capabilities. When this occurs, accomplishments of the past will seem tiny in comparison to accomplishments in the future.

Many of my columns, articles, and speeches deal with the idea of catalytic industries, innovations and trends such as the Internet of Things, flying drones, big data, driverless cars, smart homes, health tech, 3D printing, VR, swarmbots, and sensors that will be creating many of the jobs in the future.

But beyond today's seedling industries are any number of human endeavors capable of creating entire new playgrounds for business, industry, and human accomplishment.

Stepping into this topic further, I've framed my thinking around the eight dimensions for expanding the bubble of human existence.

For those of you who think three-dimensionally, expanding our bubble is like pushing on every facet of an octahedron, growing the size, reach, and capabilities in each of the X, Y, and Z axes.

The labels I've assigned to each of these dimensions include the following, and I'll explain them in more detail below:

1. **Honorability**
2. **Awareness**
3. **Purpose**
4. **Mastery**
5. **Reach**
6. **Potential**
7. **Durability**
8. **Freedom**

Why Humans?

Are humans really destined to master the universe? If so, what have we done to deserve this esteemed position?

There are many who would say the world would be a far better place without people.

If we started making a list of all the negative attributes humans possess, it would begin with words like dirty, dangerous, self-centered, moody, greedy, unreliable, hateful, destructive, self-centered, and perhaps ten thousand other descriptors that paint a very dim picture of who we are and what we've become.

For this reason, I'd like to propose the first dimension for expanding human existence—"the honorable human."

1.) The Honorable Human

Before we can ever be entrusted to receive the venerable keys to the universe, we must first prove we're worthy of this grand undertaking?

While we have achieved great things in the past, the mysteries that remain locked "behind door number three," will make our cumulative achievements to date appear as the tip of a needle in a universe filled with an endless supply of needles.

What constitutes an honorable human?

Is an honorable human someone with great integrity, loyalty, and trustworthiness you can always count on to do the right thing? Is it perhaps an evolved form of the transhuman that will arise from the singularity? Will it be a form of machine intelligence that enables us to take the higher road in every adversarial situation?

How will we ever know what attributes a person or persons will need to be deserving of this privilege? Does it have to be all of mankind or can it just be a select few?

The irony is the people we are most likely to entrust with our future are those with great courage, strength, ethics, and willingness to tackle life's greatest challenges. However, in today's world, one person's greatest hero is often someone else's greatest enemy.

We find ourselves divided by righteous differences, and these differences can lead to some very destructive consequences.

Righteous destruction is still destruction.

Similarly, a righteous conflict, battle, or killing is still a conflict, battle, or killing. Does an evil act that comes from good intentions somehow nullify the results?

At the same time, will we ever value someone without strength, conviction, drive, and passion? Probably not.

For this reason, our quest for expanding the bubble of human existence begins with a still indefinable goal of unlocking the honorable human in each of us.

What is it? What will these look like? How do we get there?

2.) Extending Human Awareness

In 1998, a column I wrote for *The Futurist* magazine took issue with the state of computer displays. Viewing the vast and growing Internet through a little square box on our desk was, in my opinion, the equivalent of watching a baseball game through a knothole.

As a solution, I proposed we experiment with a variety of different shapes for displays, starting with my favorite: a spherical display, well suited for viewing global activities such as travel itineraries, animal migrations, pollution flows, and weather patterns.

Even today, fifteen years later, we still find ourselves viewing the online world with primitive 2-dimensional flat displays. So when I heard about one satellite company's vision for developing a real-time globe, with up to the minute live video feeds of virtually every square inch on earth, naturally it caught my attention.

It wasn't just the spherical displays or video feeds of the earth that peaked my imagination, but the overall convergence of data. The number of sensory devices monitoring the earth is about to explode, and it occurred to me that a cross-pollination of data flows would radically alter our way of life.

- **Satellites monitoring the earth will grow from thousands to millions.**
- **Embedded sensors will grow from billions to trillions.**
- **Street cams, smartphones, wearables, and other connected "things" will grow from billions to trillions.**
- **The amount of data generated will burst from petabytes, to exabytes, to zettabytes, to yottabytes.**

Our growing number of data-generating devices will vividly increase awareness of the world around us. Increased awareness improves our ability to predict, and superior predictability will lead to greater control. Super awareness gives us the ability to pinpoint critical inflection points, and make changes before something serious happens.

3.) Extending Human Purpose

We are born as a baby, struggle our entire life with everything from finding food to eat, homes to live in, educating ourselves to gain more understanding, staying healthy, making friends and relationships, raising a family, earning a living, and then we die.

If we have more accomplishments in life, earn more money, have more friends, raise a bigger family, and somehow do everything better than anyone else, we will still eventually die. Right?

In a world teaming with 8.7 million different life forms, how do humans fit in?

Every past civilization, with their manmade structures, machines, systems, and cultures, has eventually succumbed to Mother Nature. Plants, animals, bacteria, and fungi have methodically removed every trace of what they left behind.

Are human accomplishments just a stepping-stone to what comes next?

We live in a world driven by prerequisites. A machinist needs to understand a single-point lathe operation before he or she can advance to multi-axial milling. Engineers need to understand the concepts of mechanical stress and strain before they start bending a cantilever beam. Metallurgists need to understand thermodynamics before they attempt phase transformations in solids. Physicists need to understand quantum mechanics before they can understand a standard model for particle physics. Mathematicians need to understand nonlinear differential equations before they can understand strange attractors.

Are all our accomplishments just stepping-stones to something else we don't know or understand yet?

Does the fact that we can ask questions like these, ponder the unponderable, think the unthinkable, and accomplish things no other species can accomplish, somehow give us a higher purpose?

If we limit our thinking to solving past problems, we can only see a very narrow spectrum of our larger purpose.

4.) Extending Human Mastery

In my column, "In Search of Anomaly Zero," I describe how we can begin to control the forces of nature and circumvent major disasters long before they happen. Once we can detect the earliest micro change in conditions and craft a timeline for an impending disaster, we will be able to create response mechanisms capable of mitigating whatever forces are in play.

Human mastery does not only give us the abilities to master the forces of nature, but every law of physics, every human condition, and every exception to every rule.

But disasters are not inevitable. Neither are illnesses, human aging, or even death.

Can we imagine something better?

If we can do a better job of controlling the negative aspects of life, and even extend it to enriching the positive aspects, is it possible to transition from the here and now to the next plateau of human existence?

The opportunities for extending human mastery are endless, and a critical piece for extending the boundaries of human existence.

5.) Extending Human Reach

Many people think we live on an overpopulated planet. But at the same time, we also live in a very under populated universe.

The option for extending the reach of humanity throughout the universe is seemingly limitless, and yet our "reach" cannot be confined to outer space.

We also know very little about inner space, such as what lies inside our planet, inside our atoms, and inside our emotions.

In a universe that is over a trillion times greater in length than the combined distance traveled by all humans in all history, we will not overcome this challenge anytime soon.

6.) Extending Human Potential

Google's Director of Engineering, Ray Kurzweil, predicted we will reach a technological singularity by 2045.Science fiction writer Vernor Vinge is betting on 2029, a date that ironically falls on the hundredth anniversary of the greatest stock market collapse in human history.

But where the 1929 crash catapulted us backwards into a more primitive form of human chaos, the singularity promises to catapult us forward into a future form of human enlightenment.

Cloaked in an air of malleable mystery, Hollywood has taken license to cast the singularity as everything from the ultimate boogeyman to the savior of humanity.

In 2013, consumer genomics company 23andMe received a patent for a designer baby kit that would allow parents to pick and choose attributes for their soon-to-be-conceived kids. This was prior to the FDA cracking down on the claims they were making.

But they were not the first. The Fertility Institutes' clinic in Los Angeles delivered the first designer baby back in 2009.

Designer babies have long been a cocktail party discussion topic with the understanding that the era of "super babies" will soon be upon us, with the prospects of creating bigger, faster, stronger humans.

Will these so-called super-babies grow up to become super-humans?

People such as Vernor Vinge and Ray Kurzweil have begun focusing in on the exponential growth of artificial intelligence, as a Moore's Law type of advancement. This has led to an entire new field of study called transhumanism, with many speculating on the next iteration of humankind and how it will be exponentially more advanced than people today.

What are the true limits to human potential, and how will we ever know if we've reached the limit?

7.) Extending Human Durability
No person should ever die... EVER! Is that our goal?

There are many reasons people die, yet these reasons may all disappear as we develop fixes and cures for everything that ails us.

Aging is currently our biggest problem. Over time we'll likely be able to fix the aging problem and delay aging indefinitely.

Injuries and disease are also problems. Over time we will likely be able to prevent and fix the issues associated with injuries and disease as well.

In columns and speeches, I have posed the question, "How long before I can 3D print a replacement body for myself?"

With major strides being made in the area of bio-printing, this becomes a legitimate question. At the same time, we still live in a very primitive time when it comes to advances on the medical front.

Perhaps the most perplexing problem to fix will be that of deviant behavior, because the idea of fixing deviant behavior presumes we will have a good way of sorting out the dividing line between deviant and non-deviant behavior. But there again, over time we will likely develop medical or behavioral strategies that address deviant behavior.

If we have the ability to fix the problems involved with aging, injury, disease, and deviant behavior, theoretically we can create a society of people capable of living forever.

Is that our goal? And if not, why not?

8.) Extending Human Freedom

For many of us, the idea of freedom conjures up the symbols of containment, like steel shackles or doors that are somehow unlocked before us, allowing us to breathe the rare air of independence.

But beyond the insular notion of conscious confinement is a life unrestrained by the bonds of our own limitations.

Universal freedom comes with the sense that anything is possible.

If people did not have to worry about illness, safety, natural disasters, the limitations of time and space, and human frailty, what will then be possible? What can they then achieve?

How long before we have the unbridled freedom to live life on a macro level, take on projects larger than our solar system, and begin living outside our own bubble?

Labeling these Dimensions

I started this chapter talking about how we're still trapped in the bubble of human existence, but finding a way to expand our bubble, or actually live beyond our reality sphere is a challenging big picture perspective.

Granted, we've been doing it all along, first by taking micro steps, but moving to giant leaps over the past century.

What I'm suggesting here, by adding labels to each of these dimensions, is that this is our calling, our "unfinishable mandate" to continually stretch, grow, propagate, and master not only the world around us, but also the entire universe.

The human race is genetically predispositioned to push the envelope, color outside the lines, and reach for things that will forever be unreachable.

As individuals, there will always be some who are content to find inner peace and live a minimalist lifestyle. But as a species, we will always be driven by a need to make a difference, be admired for our accomplishments, and create moments of triumph in our otherwise pale existence.

We have taken only the first step in a trillion-mile journey. The next few steps, in my opinion, will be absolutely amazing.

Rewriting the Prime Directive

Rather than living in a world with people fighting people, the true battles that lie ahead will test us on every conceivable level. On the grandest of scales, we will find ourselves confronted with forces larger than our entire solar system, and on the tiniest of scales, nanotechnology and sub-atomic particles will confound us with challenges we never dreamed could exist. These battles will require far more than brilliant minds, personal tenacity, and military might.

People of tomorrow will need to be prepared for a higher calling. This higher calling will be to pre-empt crises before they occur, anticipate disasters before they happen, and solve some of mankind's greatest problems, starting with the problem of our own ignorance.

Much like a person walking through a dark forest with a flashlight that illuminates but a short distance ahead, each step forward gives us a new perspective by adding light to what was previously dark. *The people of tomorrow will simply need a better flashlight.*

Until now, ours has been a dance with the ordinary. History shows us we are immersed in cycles, systems, and patterns that repeat again and again. Tomorrow's history books will show us all patterns are made to be broken, all cycles waiting to be transformed.

Higher education will need to position itself on the bleeding edge of what comes next.

We will always need systems for looking backwards to understand where we have come from, but a new breed of visionaries, bestowed with unusual tools for preempting disasters, are destined to become our most esteemed professionals.

Life in the future will not be easy, nor should it be.

Perhaps a simpler way of stating our Prime Directive would make clear the size of the opportunity—and the challenge—awaiting us:

"Preparing humanity for worlds unknown, preparing our minds for thoughts unthinkable, and preparing our resolve for struggles unimaginable."

There is no larger, nor more vital, task today—or tomorrow. I look forward to joining you on this quest.

ACKNOWLEDGEMENTS

The people I've surrounded myself with have had a huge impact on where this journey has taken me, and I'm honored to be surrounded by some of the best and brightest in the world.

During the past few months, one person in particular, Keith Ferrell, former Editor-in-chief for Omni Magazine, has served as my editorial mentor, coach, advisor, word-smith, teacher, confidant, and friend as we worked our way through the creation of this book.

The DaVinci Institute has attracted a considerable following since 1997 when it was first formed, and I now find myself in the company of remarkable people leading remarkable lives.

To the Senior Fellows who serve as the elite brain trust for the Institute, I offer my sincerest thanks and appreciation.

To those of you who have participated in the Futurist Mastermind Groups and the Mad Scientist Club, I will forever be indebted for your brilliance.

To Michael Cushman, our director and lead consultant for the Vizionarium Project (Corporate Consulting Division), and close personal friend, your way of thinking about the world has influenced innumerable aspects of my personal and professional life.

To my "secret agent," trusted-assistant, and all-around handler-of-my-speaking-schedule, Micale Medina, you've been an incredible help. Your middle-of-the-night-emails in response to my middle-of-the-night-emails help keep me motivated and put a smile on my face.

To our staff at the Institute—Jan Wagner, Mike Morgan, Jessica Morgan, Kaitlyn Maddox, Jennifer Alberts, Sandy Kriete and all of our volunteers—you make the work we do seem easy and fun. To you my sincerest thanks and appreciation.

To our instructors for DaVinci Coders—Jason Noble, Julien Lynge, Caron Newman, Books Hollar, Mike Cullerton, and Ben Eveloff—you've helped build a highly respected school, a Micro College, and critical foundation for the operation of the DaVinci Institute.

ABOUT THOMAS FREY

Over the past decade, Thomas Frey has built an enormous following around the world based on his ability to develop accurate visions of the future and describe the opportunities ahead.

Having started seventeen businesses, himself and assisting on the development of hundreds more, the understanding he brings to his audiences is a rare blend of reality-based thinking coupled with a clear-headed visualization of the world ahead.

Predicting the future has little value without understanding the driving forces behind the trends, subtle nuances that can be leveraged, and implications for both the people directly affected in the industry as well as others farther down the technological food chain.

As part of the celebrity speaking circuit, Tom continually pushes the envelope of understanding, creating fascinating images of the world to come. His keynote talks on futurist topics have captivated people ranging from high level government officials to executives in Fortune 500 companies including NASA, IBM, Disney, Federal Reserve Bank, KPMG, Pepsico, Caterpillar, Unilever, Frito Lay, Social Security Administration, GE, Toshiba, California State University, American Library Association, Hunter Douglas, International Council of Shopping Centers, National Association of Federal Credit Unions, Times of India, and many more.

Because of his work inspiring innovators, the Denver Post and Seattle Post Intelligencer have referred to him as the "Dean of Futurists".

Before launching the DaVinci Institute, Tom spent 15 years at IBM as an engineer and designer where he received over 270 awards, more than any other IBM engineer. He is also a past member of the Triple Nine Society (High I.Q. society over 99.9 percentile).

Thomas has been featured in thousands of articles for both national and international publications including *New York Times, Wall Street Journal, U.S. News & World Report, Wired Magazine, Fast Company, Forbes, National Geographic, USA Today, Times of India*, and virtually every TV station in North America and Australia. He currently writes a weekly "Future Trend Report" newsletter and a weekly column for FuturistSpeaker.com.

CONTACT PAGE

"Secret Agent" Micale Medina

Those wishing to contact Futurist Thomas Frey either to book a speaking event, press interview, consulting inquiry, or for any other reason should work with "Secret Agent" Micale Medina. Micale can be reached at 303-666-4133 or Micale@DaVinciInstitute.com.

Anyone wishing to take a tour of the DaVinci Institute should work with Mikael Morgan. He can be reached at 303-666-4133 or Mikael@DaVinciInstitute.com.

DaVinci Institute, 9191 Sheridan Blvd, Suite 300, Westminster, CO 80031

A free eBook edition is available with the purchase of this book.

To claim your free eBook edition:

1. Download the Shelfie app.
2. Write your name in upper case in the box.
3. Use the Shelfie app to submit a photo.
4. Download your eBook to any device.

Shelfie

A free eBook edition is available
with the purchase of this print book.

CLEARLY PRINT YOUR NAME ABOVE IN UPPER CASE

Instructions to claim your free eBook edition:
1. Download the Shelfie app for Android or iOS
2. Write your name in **UPPER CASE** above
3. Use the Shelfie app to submit a photo
4. Download your eBook to any device

Print & Digital Together Forever.

Snap a photo

Free eBook

Read anywhere

The Morgan James
Speakers Group

↗ www.TheMorganJamesSpeakersGroup.com

We connect Morgan James published
authors with live and online events
and audiences whom will benefit
from their expertise.

CPSIA information can be obtained
at www.ICGtesting.com
Printed in the USA
BVHW03s1129040818
523242BV00003B/17/P

9 781683 500179